Advanced LabVIEW® Labs

Advanced LabVIEW® Labs

John Essick

Reed College
Physics Department

Prentice Hall
Upper Saddle River, New Jersey 07458

Library of Congress Cataloging-in-Publication Data

Essick, John
  Advanced Labview Labs / John Essick
    p.   cm.
  ISBN 0-13-833949-X
    1. LabVIEW.   2. Science—Experiments—Data processing.
3. Scientific apparatus and instruments—Computer simulation.
4. Computer graphics.   I. Title.
Q185.E69 1998   98-40830
507.8—dc21   CIP

Executive Editor: Alison Reeves
Editorial Assistant: Gillian Kieff
Executive Managing Editor: Kathleen Schiaparelli
Assistant Managing Editor: Lisa Kinne
Art Director: Jayne Conte
Cover Designer: Bruce Kenselaar
Manufacturing Manager: Trudy Pisciotti
Production Supervision/Composition: WestWords, Inc.

ISBN 0-13-833949-X

Prentice-Hall International (UK) Limited, *London*
Prentice-Hall of Australia Pty. Limited, *Sydney*
Prentice-Hall Canada Inc., *Toronto*
Prentice-Hall Hispanoamericana, S.A., *Mexico*
Prentice-Hall of India Private Limited, *New Delhi*
Prentice-Hall of Japan, Inc., *Tokyo*
Prentice-Hall (Singapore) Asia Pte. Ltd., *Singapore*
Editora Prentice-Hall do Brasil, Ltda., *Rio de Janeiro*
Prentice-Hall, Inc., Upper Saddle River, New Jersey

To my wife, Katie

Contents

Preface

The personal computer now stands alongside such equipment as the multimeter and the oscilloscope as one of the standard tools of scientific research. This powerful instrument is included in most modern-day laboratories because it facilitates sophisticated measurements beyond the reach of manually operated set-ups and is invaluable in analyzing the resulting data. Thus, most instructors involved in the design of upper-division science and engineering curricula feel strongly that a portion of each student's advanced laboratory experience should be devoted to instruction in computer-based experimental skills.

Quite commonly, the topic of computer-based experimentation is presented in college courses through a sequence of experiments that explore the low-level details of the computer interface. Students investigate the mechanisms by which analog-to-digital and digital-to-analog operations are accomplished at the chip level, then go on to study the machine-level design of microprocessors. Finally, students learn to program the microprocessor chip using assembly language instructions and write a routine to carry out a primitive laboratory task. With this firm grounding in basic computer operation at the machine level, it is hoped that the student, when actually carrying out a computer-based laboratory experiment at a future date, will have the basis to understand pertinent issues and to troubleshoot problems.

However, in response to the fast-paced advances that have occurred in scientific instrumentation over the past 15 years, many educators have begun to revise their presentation of computer-based research skills. With the low cost, sophistication, and reliability of commercially available products, most active researchers no longer construct homemade data-taking hardware or write assembly language code. Rather, they wisely purchase robust data acquisition systems that can be controlled with symbolic ("high-level") commands. In contrast to the intimate knowledge of chip-level processes required in the past, today's researcher is routinely called upon to implement a wide range of high-level computer skills—for example, using a symbolic programming language to control a data acquisition process, storing the resultant data in a convenient form within a computer's filing system, communicating with instruments through the General Purpose Interface Bus (GPIB), or writing a program to analyze data in real time using sophisticated techniques such as the Fast Fourier Transform. This insight opens the door to a new realm of possibilities in the design of an advanced instructional laboratory. Rather than devoting the entire course to the time consuming task of teaching the details of data-taking technology, an instructor is freed to contemplate a much broader range of instruction in which students learn to use the computer for instrument control, data gathering, analysis, and interpretation. With the proper mix of offerings, such a course has the potential to outfit students with a repository of relevant ideas, tools, and experiences that they can directly call upon in their first real research projects, and in the years beyond.

Inspired by such musings, while teaching at Occidental College eight years ago, I began to include a high-level presentation of computer-based experimentation in my upper-division laboratory course. Using National Instruments GPIB boards plugged into the expansion slots of several personal computers and the related language-specific driver software, I sought to teach students how to computer control the GPIB-interfaced instrumentation they were using to perform solid-state physics experiments. While this pedagogical experiment worked very well for some students, I judged it a failure for others, and it was not difficult to identify the source of this varying success. Those students with a solid background in computer programming could "hit the ground running" and almost immediately start to explore my intended topic—computer-based experimentation. Of course, these computer-literate students "spoke" a variety of programming languages, but I solved this minor inconvenience by developing a library of BASIC, FORTRAN, and C compilers and driver software for the course. The real problem was students with little or no computer programming background. These students were confronted with the hurdle of learning at least the rudiments of a particular computer language prior to tackling the real task at hand—computer interfacing. While most of the computer neophytes were fascinated by the prospect of controlling an instrument remotely, it proved impossible for them, given the 10-week time period of my course, to become competent in a language such as C at the level required to code a data acquisition program. I realized that I had to remove this software stumbling block so that each of my future students could be successful, but I never satisfactorily solved this teaching problem while at Occidental.

I first saw a LabVIEW system when I arrived at Reed College five years ago. After playing with this graphical programming language for a few hours, I became convinced that it was the perfect environment in which to teach computer-based research skills. Its intuitive use of wired-together icons to perform simple-to-understand tasks made adding two numbers, building an array, parsing a string, and digitizing an analog voltage practically equivalent in programming ease. Moreover, with a minimum of difficulty, one could implement all sorts of sophisticated analysis techniques such as curve fitting, digital filtering and fast Fourier transforming. On top of all that, LabVIEW programming was fun! It seemed to me that even a complete computer novice, given the proper systematic instruction, could learn this language in a short amount of time and then use it to perform a wide variety of laboratory tasks using the computer.

Five years later, I can say that LabVIEW has far exceeded my initial optimistic expectation as an effective pedagogical tool. Its programming features are clear, coherent, powerful, comprehensive and entertaining, enabling an instructional presentation of computer-based experimentation in which students create meaningful programs that illustrate useful concepts at each step of the learning curve. LabVIEW programs are modular, so that after each is created and understood, it becomes part of a library that can be used later as a building block of a more sophisticated program. With just a few weeks of training, students develop the ability to explore their own ideas by modifying programs or by writing appropriate new ones and retain these skills long after the course is over.

Advanced LabVIEW Labs contains the collection of LabVIEW-based exercises that I have developed and used successfully to present computer-based experimentation in my junior-level Advanced Laboratory course at Reed. The first seven chapters are designed to teach the LabVIEW programming language including its four control structures (While Loop, For Loop, Case Structure, and Sequence Structure), three graphing modes (Waveform Chart, Waveform Graph, and XY Graph) and File I/O. Chapters 8 and 9 explore two interesting and commonly used analysis techniques— least-squares curve fitting, and the Fast Fourier Transform. Chapters 10 and 11 investigate the analog-to-digital and digital-to-analog conversion functions available on a National Instruments data acquisition board. Finally, Chapter 12 provides instruction in LabVIEW-based GPIB control of laboratory instrumentation. Based on my experience in using these materials for instruction at Reed, each chapter takes approximately two to three hours for a student to complete.

Advanced LabVIEW Labs is very much a hands-on presentation of its subject. In a sense, the book is one long exercise through which students work their way. Because LabVIEW is a graphical (as opposed to text-based) programming language, its concepts and programs can be naturally rendered in pictures. Most pages of the book contain a screen shot (or two) of the LabVIEW program students are asked to reproduce. In earlier chapters, as the programming language is being taught, detailed instructions and multiple screen shots of intermediate programming steps are given to assist the student in successfully completing the exercises, while in later chapters, students are expected to complete the programs with much less coaching.

A modular approach to problem solving and programming is emphasized throughout the text. LabVIEW programs written in earlier chapters are used as sub-programs within the more advanced programs of later chapters. For example, the Sine Wave program written in Chapter 3 is used in the Fast Fourier Transform program developed in Chapter 9, which is then used as a sub-program within the Spectrum Analyzer built in Chapter 10.

Similarly, the construction of a temperature control system provides a unifying thread through several of the chapters. The calibration curve for a temperature-sensing thermistor is determined in Chapter 8, the temperature measurement set-up is assembled in Chapter 10, and the complete temperature control system is constructed as a somewhat open-ended project in Chapter 11. Attention-grabbing and useful electronic components such as a thermistor and thermoelectric device are used in this project to enhance student interest. Appendix I provides detailed construction plans for the low-cost (approximately $80) temperature control apparatus needed.

All of the programs presented in this book are written using LabVIEW Version 5. The text will also work perfectly well for Version 4 users. However, LabVIEW 3 owners are gently advised that it might be time to consider an upgrade. While Version 3 of LabVIEW possesses all the capabilities necessary for this book, the editing steps are different (and primitive) compared to the later versions. But if you don't mind the extra transcription work, all of the *Advanced LabVIEW Labs* exercises can be executed in LabVIEW 3.

To aid readers in completing the exercises, I use the following conventions throughout the book: **Bold** text designates the features such as graphical icons, palettes, pull-down menus, and menu selections that are to be manipulated in the course of constructing a LabVIEW program. The descriptive names that label controls, indicators, custom-made icons, programs, disk files, and directories (or folders) are given in the **straight** font. *Italic* text highlights character strings that the programmer must enter using the keyboard and also signals the first-time use of important terms and concepts.

Advanced LabVIEW Labs is compatible with both the Full Development System and the Student Edition of LabVIEW. To minimize the amount of valuable laboratory time devoted to learning the LabVIEW programming language, an instructor might consider having students purchase personal copies of the low-cost Student Edition software and then perform the software-only Chapters 1 through 9 as homework on their own computers.

The following table lists the items required to work the exercises in this book.

Chapter	Additional Software	Plug-In Board	Additional Hardware
1			
2			
3			
4	Word Processor[a] and/or Spreadsheet Analysis[b] Program		
5			
6			
7			
8			
9			
10		DAQ[c]	Thermistor[e], Constant Current Circuit[e], Function Generator
11		DAQ[c]	Thermistor + Constant Current Circuit, Thermoelectric Device + Current Driver Circuit[f]
12		GPIB[d]	GPIB Interfaced Instrument[g]

[a]Such as Microsoft Word
[b]Such as Microsoft Excel
[c]Such as E Series MIO, PCI-1200, Lab-PC or Lab-NB Board
[d]Such as AT-GPIB, PCI-GPIB, NB-GPIB
[e]See Chapter 10
[f]See Appendix I
[g]Such as HP34401A Multimeter

Some familiarity with op-amps is necessary to understand the circuits in Chapter 10 and 11. In my year-long Advanced Laboratory course, students acquire this requisite background from seven weeks of advanced electronics experiments (amplifying, filtering, timing and logic circuits) that precedes their instruction in computer-based data acquisition and analysis. The material in *Advanced LabVIEW Labs* is then covered over the course of seven weeks in the following manner: Chapters 1–3 (Week 1),

Chapters 4–7 (Week 2), Chapters 8–9 (Week 3), Chapter 10 (Week 4), Chapter 11 (Week 5), Chapter 12 (Week 6 and 7). A one-hour lecture each Monday provides students with an overview of that coming week's lab work. After completing the LabVIEW–based instruction, my Advanced Laboratory students then devote the remaining 11 weeks of the course to open-ended investigations of physical phenomena.

One indication of the effectiveness of *Advanced LabVIEW Labs* is that many Reed students have successfully utilized their LabVIEW programming expertise to execute sophisticated open-ended Advanced Laboratory experiments during the latter portion of the course. Many also have gone on to use LabVIEW in their Senior Thesis projects. LabVIEW programming, because of its wide use in industrial and research labs, has also proved to be a marketable skill to students applying for summer internships, full-time jobs, and graduate schools. I hope you, too, will find *Advanced LabVIEW Labs* to be an interesting and beneficial introduction to the power and flexibility of LabVIEW-based data acquisition and analysis.

Any suggestions or corrections are gladly welcomed and can be sent to John Essick, Reed College, 3203 SE Woodstock Blvd., Portland, OR 97202 USA or jessick@reed.edu.

For their advice and assistance while I prepared this book, I would like to thank my colleagues David Griffiths, Richard Crandall, Mary James, and Mark Beck, my students Zach Nobel and Ben Palmer as well as Ravi Marawar, Mahesh Chugani and Lisa Wells of National Instruments. Also I'd like to express my appreciation to Roger Bengston (University of Texas, Austin), Grant Hart (Brigham Young University), Robert B. Muir (University of North Carolina, Chapel Hill), Edward Nadgorny (Michigan Technological University), and Ronald Ransome (Rutgers University) for their helpful reviews of the manuscript. Special thanks to Alison Reeves and Gillian Kieff of Prentice Hall, and Jennifer Maughan and her crew at WestWords, for making this book possible. I gratefully acknowledge the support of the M.J. Murdock Charitable Trust and the National Science Foundation in purchasing LabVIEW systems for the Reed College Physics Department. Finally, to my family: Thank you for being lively and loving while I worked on this project.

John Essick
Portland, Oregon

CHAPTER 1

The While Loop And Waveform Chart

Welcome to the world of LabVIEW, an innovative graphical programming system designed to facilitate computer-controlled data acquisition and analysis. In this world, you—the LabVIEW user—will be operating in a programming environment that is different from that offered by most other programming systems (for example, the C and BASIC computer languages). Rather than creating programs by writing lines of text-based statements, in LabVIEW you will code programming ideas by selecting and then properly patterning a collection of graphical icons.

A LabVIEW program consists of two windows, the *front panel* and the *block diagram.* Once you have completed a program, the front panel appears as the face of a laboratory instrument with your own design of knobs, switches, meters, graphs, and/or strip charts. The front panel is the program's user-interface, that is, it facilitates the interaction of supplying inputs to and observing outputs from the program as it runs.

The block diagram is the actual LabVIEW programming code. Here resides the graphical images that you have selected from LabVIEW's well-stocked libraries of icons. Each icon represents a block of underlying executable code that performs a particular useful function. Your programming task is to make the proper connections between these icons using a process called *wiring,* so that data flows amongst the graphical images to accomplish a desired purpose. Because the icon libraries are designed specifically with the needs of scientists and engineers in mind, LabVIEW enables you—the modern-day experimentalist—to write programs that perform all of the laboratory tasks required for your state-of-the-art research including instrument control, data acquisition, data analysis, data presentation, and data storage.

To begin developing your skill in graphical programming, in this first chapter I will guide you through the steps of creating a LabVIEW program. Together (in later chapters, as your LabVIEW expertise increases, you will work your way through the exercises much more independently), we will write a program that systematically increments the argument of the sine function and plots the resulting sine wave as time progresses. In the process of writing this program, you will learn about the *While Loop,* one of LabVIEW's four available program control structures, as well as one of LabVIEW's three graphing modes, the *Waveform Chart.*

SINE WAVE GENERATOR USING A WHILE LOOP AND WAVEFORM CHART

Find the LabVIEW program (which will be in a directory or folder named LabVIEW, LabVIEW Student Edition, or LVSE, depending on your particular computing system) and launch it by double clicking on its icon or name. If you are using one of the latest version of LabVIEW, you will be presented with a dialog box on which you should select the **New VI** option.

After a few moments, an untitled front panel will appear in the foreground of your screen with a slightly offset block diagram in the background. In addition, a menu bar for an array of pull-down menus that contain LabVIEW program editing items will appear at the top of your screen (for a Macintosh-based system, as shown below) or at the top of the window (in the Windows environment).

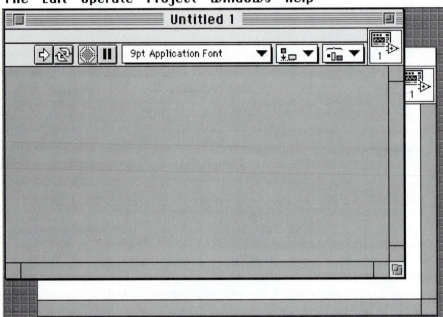

There are three ways to toggle the block diagram between the background and foreground:

- Select **Show Diagram** from the **Windows** pull-down menu.
- Click your mouse pointer on a visible region of the block diagram while it is in the background.
- Use the keyboard shortcut by typing *<Cmd><E>* on a Macintosh or *<Ctrl><E>* on a Windows system.

Practice toggling the block diagram between the foreground and background using each of the three available methods.

Block Diagram Editing

To begin programming, place the blank block diagram in the foreground. The objects to be placed on this block diagram as you write your LabVIEW program are found in a repository called the *Functions Palette.* Activate the Functions Palette by selecting **Show Functions Palette** from the **Windows** menu.

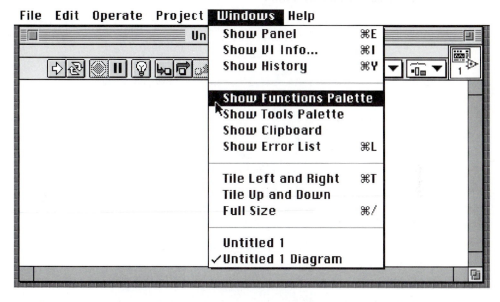

A "floating" Functions Palette will appear on your screen, as shown below, which can be placed in a convenient spot by clicking and dragging its title bar.

The Functions Palette is an organized collection of subpalettes, where each subpalette contains a group of related programming objects. Introduce yourself to the contents of the Functions Palette by passing the mouse cursor over each subpalette button and noting each subpalette name as it appears at the top of the window.

Place the cursor over the **Structures** subpalette button, then click and hold down your mouse button. The **Structures** subpalette will become visible as shown.

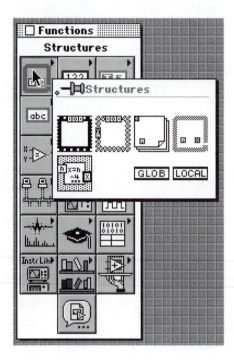

While keeping the mouse button depressed, move the cursor over the objects in the **Structures** subpalette and watch their names appear at the top of the window. Then select the **While Loop** structure from the subpalette by releasing the mouse button after placing the cursor over this object. Once you've made this selection, the cursor will appear as a miniature of the While Loop structure when it is placed within the block diagram window. To place a While Loop within the diagram, click where you want the upper left corner of the loop to be, then, while holding down the mouse button, drag the cursor to define the size of your loop.

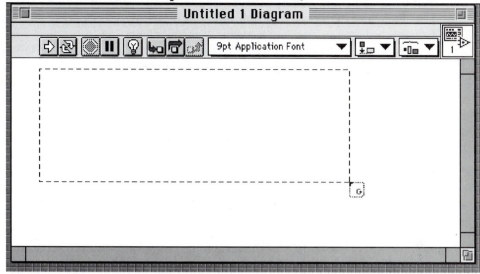

When you release the mouse button, the While Loop will appear as shown:

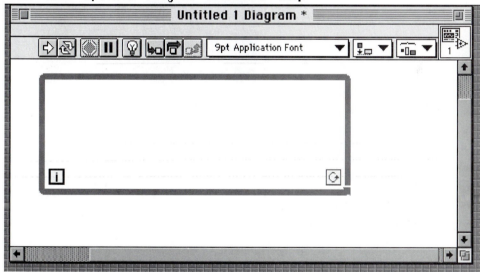

If you are dissatisfied with the dimensions of your While Loop, this can be remedied by using the *Positioning Tool* ⤓ , whose job is to select, move, and resize objects. The ⤓ is one of several available LabVIEW editing tools displayed in the *Tools Palette,* which may or may not be visible at the moment.

Activate the Tools Palette by selecting **Show Tools Palette** in the **Windows** menu. The "floating" Tools Palette will appear as shown below and, like the Functions Palette, can be relocated to a convenient location by clicking and dragging its title bar.

The Positioning Tool may already be selected on your Tools Palette, in the manner shown above. If not, use the mouse cursor to select it now. A short description of each of the tools can be obtained from the *tip strip* that appears when placing the cursor over each tool's button. An example of a tip strip is shown here.

The Positioning Tool can be used to resize your While Loop by the following procedure. Place the ⤓ at one of the loop's corners. At the corner, the ⤓ will transform into a resizing handle ⌐. Click and drag this cursor to redefine the dimensions of your While Loop. When you release the mouse button, the While Loop of desired size will appear.

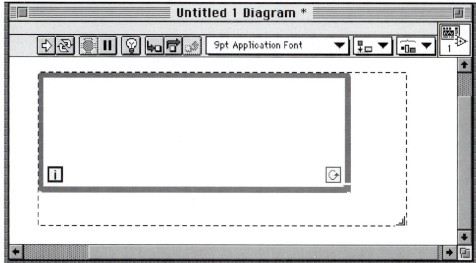

The While Loop structure is used to control repetitive operations. It will repeat-edly execute the subprogram written within its borders (called the *subdiagram*) until a specified Boolean value is no longer TRUE. Thus this structure is equivalent to the following code:

```
Do
     Execute subprogram (which sets condition)
While condition is TRUE
```

Within the While Loop, you will find the *conditional terminal* and the *iteration terminal*. At the end of each loop iteration, LabVIEW checks the value of. If it is TRUE, the value of is incremented by one, and the loop begins another execution. The initial value of is zero and an unconnected has a default value of FALSE. Thus, in a While Loop where is left unconnected to anything, the loop will execute exactly once and the final value of will be zero. In this chapter, we will cause a While Loop to execute numerous iterations by connecting to a Boolean control terminal that is usually TRUE.

Now, let's write a program that will generate and plot a sine wave. With a While Loop already on your block diagram, select the **Sine & Cosine** icon from the Functions Palette. To find this icon, first select the **Numeric** subpalette, then the **Trigonometric** sub-subpalette. From now on, such a sequence of choices will be indicated as follows: **Functions>>Numeric>>Trigonometric**. For the coming exercises, you can just as well use the **Sine** icon, also found in **Functions>>Numeric>>Trigonometric**. The **Sine &**

Cosine is more thrilling, though, because it will allow you the experience of wiring to an icon with multiple outputs.

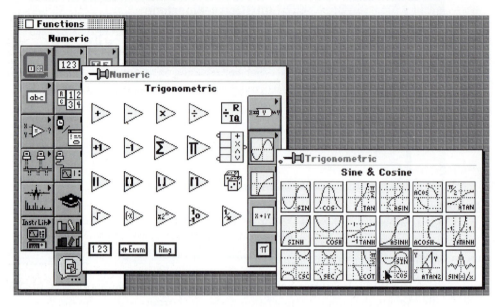

Once you've selected the **Sine & Cosine** icon from the Functions Palette by releasing the mouse button, place the cursor at the location within your block diagram at which you wish the icon to appear.

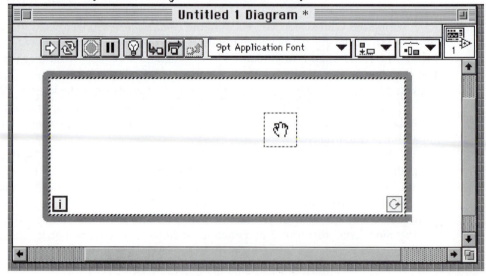

Then click the mouse and Whoomp!—there it is.

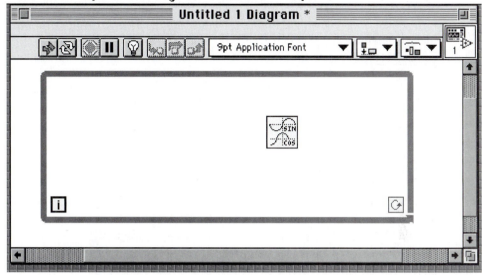

If you subsequently decide that you'd like to move this icon to some other location, you can do so through use of the Positioning Tool. With the ⬉ , click on the **Sine & Cosine** icon. When thus selected, the icon will become highlighted by a moving, dashed border called a *marquee*. With the mouse button depressed, drag the highlighted icon to the newly desired location within your While Loop. Once properly placed, release the mouse button and move the cursor to an empty spot on the block diagram. Then click the mouse to deselect the icon.

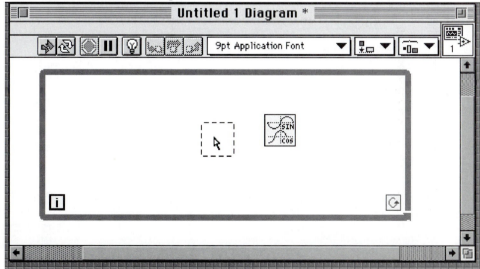

Once an object is highlighted with a marquee, there are a couple of other handy movement techniques. If you hold down the *<Shift>* key and then drag the object, LabVIEW will only allow purely horizontal or vertical motion. Also, you can move the selected object in small, precise increments by pressing the keyboard's *<Arrow>* keys, rather than dragging with the mouse. Try moving the **Sine & Cosine** icon using both of these techniques.

The function of the **Sine & Cosine** icon is described in the following *Help Window:*

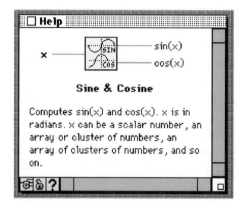

In a Help Window, the icon's input connections are shown on the left and output connections on the right. Thus we see that the **Sine & Cosine** icon accepts an argument **x** in radians and returns values for **sin (x)** and **cos (x)** at its two outputs.

You may see this Help Window for yourself by selecting **Show Help** from the **Help** menu.

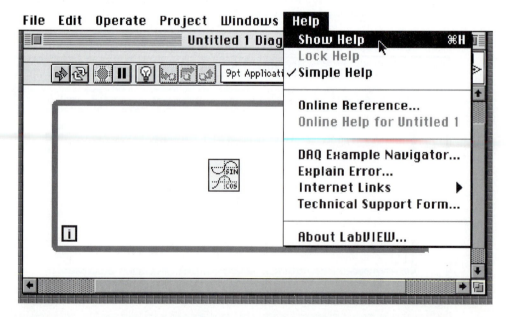

When the Help Window appears, place it in a convenient position using the ⊾ . Then place ⊾ over the **Sine & Cosine** icon to view its description.

When you have no further need for the Help Window, it can be toggled off in the **Help** menu. As an alternative to mouse control, try toggling the Help Window on and off using the following keyboard shortcut: *<Cmd><H>* (Macintosh) or *<Ctrl><H>* (Windows).

Now let's write a program that generates a sequence of points that follows the sine function. First, we need to configure the While Loop so that it will repetitively perform the operation defined within its borders. One (crude) method of accomplishing this goal is the following: Select a **Boolean Constant** from **Functions>>Boolean**:

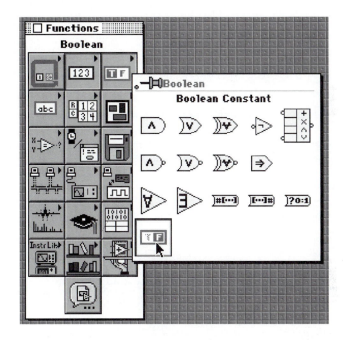

Place the **Boolean Constant** 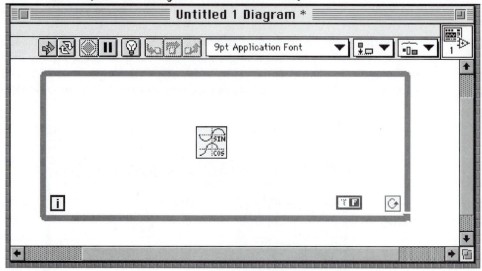 near the While Loop's conditional terminal . Note that the default value of is FALSE.

We will now connect the Boolean Constant to by using the *Wiring Tool* . Go to the Tools Palette and select the Wiring Tool by clicking on it.

Position the over the **Boolean Constant** icon, so that the begins to blink. Then click the mouse. You have now tacked down one end of a wire to the .

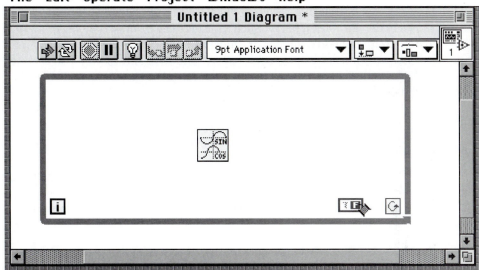

Smoothly move the Wiring Tool to the right until it is over the conditional terminal and the  is blinking. Click the mouse.

If all goes well, you will see a dotted green wire connecting the and icons. This green wire is LabVIEW's method of indicating that Boolean data is to be passed between two programming objects; in this particular case, from the Boolean Constant to the conditional terminal .

If, while using the ✶, you make a mistake and try to form an improper connection between two objects, the resulting "bad wire" will appear as a black dashed line, as shown in the illustration below. Such a mistake can be erased by selecting **Remove Bad Wires** from the **Edit** menu or, more simply, by using the keyboard shortcut: *<Cmd><B>* (Macintosh) or *<Ctrl><B>* (Windows). Alternately, you can highlight the bad wire with the Positioning Tool, then erase it by pressing the *<Delete>* key on your keyboard.

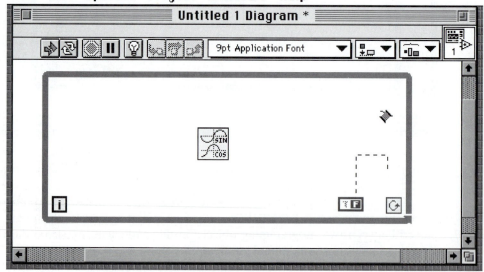

To cause the While Loop to iterate continually, we need simply to switch the Boolean Constant to the value of TRUE. This TRUE value then will always be passed to the conditional terminal at the end of each loop iteration, causing the While Loop to re-execute.

It is the job of the *Operating Tool* 🖑 to change values of constants on the block diagram. Select 🖑 on the Tools Palette, place it over the TRUE indicator of the 🔲 icon, then click. The Boolean Constant will now have its TRUE indicator highlighted 🔲, indicating that the TRUE value has been selected.

We will now use the While Loop's iteration terminal 🅸 as a source of ever-increasing argument x for our sine function. Use the ⬧ to position the 🅸 in the neighborhood of the Sine & Cosine icon, then wire the 🅸 to the Sine & Cosine's argument **x** input.

Here is a method of producing proper wiring. First, activate the Help Window by the keyboard short cut *<Cmd><H>* (Macintosh) or *<Ctrl><H>* (Windows), then position the Help Window in a convenient out-of-the-way place. Now, select the Wiring Tool from the Tools Palette, position the ✶ over the 🅸 so that this icon blinks, and click to tack down one end of a wire to the 🅸. Move the ✶ smoothly over to the terminal for the Sine & Cosine's argument **x** input, using the Help Window as your guide.

When the ❖ is positioned correctly, the terminal for the **x** input will begin to blink both on the block diagram and in the Help Window. To erase any further doubt regarding the identity of the terminal to which you are wiring, the terminal's name is supplied by a tip strip. Click the mouse to complete the wiring. As an additional wiring aid, "whiskers" appear from each input and output as the Wiring Tool closely approaches the icon. Dots terminate the whiskers at inputs, while output whiskers have no dots. These whiskers are especially helpful when wiring to icons with a multitude of inputs and outputs.

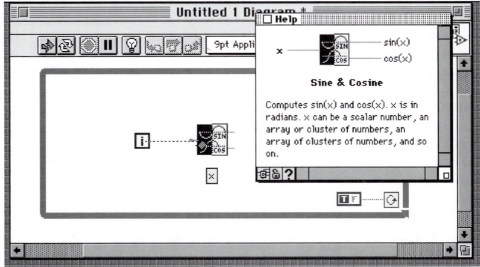

If done correctly, a colorful wire will connect the ⬛ to the **x** input, indicating that numerical data is to be passed between these two objects. Remember, if you make a mistake, the illegal wiring will result in a black dashed line that can be most easily erased by the **Remove Bad Wires** command's keyboard shortcut: *<Cmd><B>* (Macintosh) or *<Ctrl><B>* (Windows).

In LabVIEW, the color of a wire carrying numerical data indicates the manner in which the number is represented. Blue wires and icons denote integers, which come in one-, two- and four-byte varieties, while orange indicates single- (four-byte) or double-precision (eight-byte) floating point numbers. We will soon discover how to control the exact formatting of a particular numerical value. For now, simply note that the ⬛ icon and wires emanating from it are blue, indicating integer values. However, a gray *coercion dot* appears where this blue wire connects to the Sine & Cosine icon input, denoting that the icon is automatically converting this integer input into the floating-point format it requires for its argument **x** input.

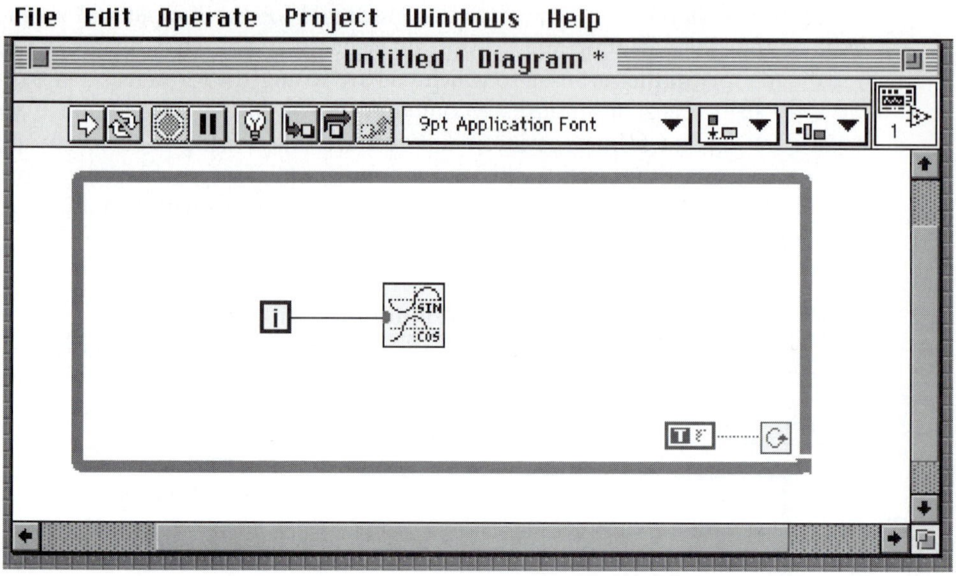

File   Edit   Operate   Project   Windows   Help

Untitled 1 Diagram *

9pt Application Font

In the above block diagram, all of the wires are straight. If, instead, you'd like to create a wire that includes right-angle bends, simply click the mouse while wiring at the desired "bend positions" and continue wiring perpendicularly.

Online Reference

To obtain more detailed information about the inner workings of a particular icon, LabVIEW's *Online Reference* resource can be accessed by clicking on the Question Mark ▣ at the bottom of the icon's Help Window or by selecting **Online Reference…** in the **Help** pull-down menu.

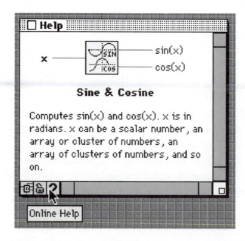

Help

x ———▭——— sin(x)
          └——— cos(x)

Sine & Cosine

Computes sin(x) and cos(x). x is in radians. x can be a scalar number, an array or cluster of numbers, an array of clusters of numbers, and so on.

Online Help

A window giving high-level comments about the icon appears first.

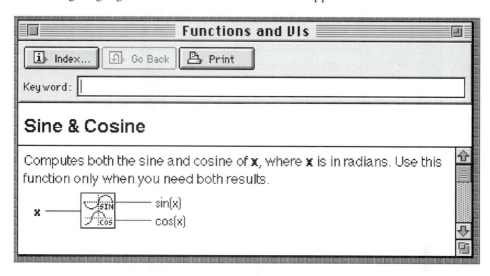

For more detailed information, place the cursor over the object of interest and click, as shown here for the **x** input.

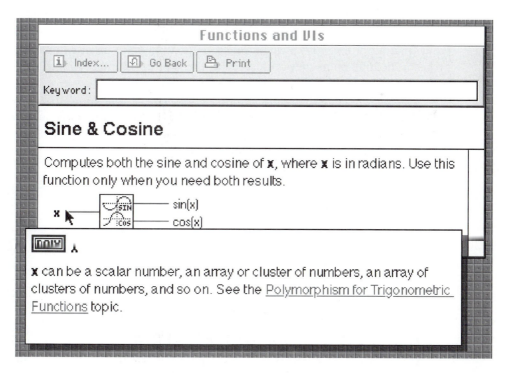

A window pops up to inform us that the **x** input is *polymorphic*. The definition of polymorphic can be found by clicking on the highlighted phrase "Polymorphism for Trigonometric Functions."

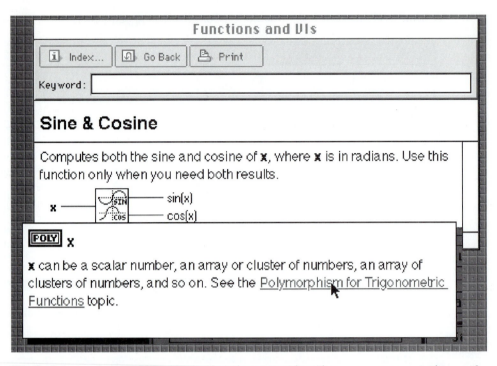

We find that polymorphism means that the output has the same representation as the input (that is, for example, a single-precision floating-point input will yield a single-precision floating-point output), except in the case of an integer input. For integer inputs, the output is a double-precision floating-point number.

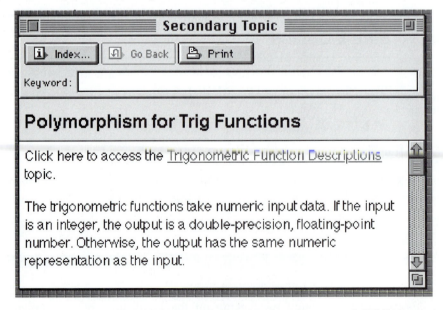

Now, let's instruct LabVIEW to plot the sine function as each While Loop iteration generates a new value. A graph falls under the generic category of user-interface, which is the domain of the front panel. Switch to the front panel by using *<Cmd><E>* (Macintosh) or *<Ctrl><E>* (Windows).

Front Panel Editing

A user typically wishes to interact with a program by supplying inputs and observing outputs. In LabVIEW, these operations are facilitated by a wide choice of knobs, switches, meters, and graphs found in the *Controls Palette*. Activate the Controls Palette by selecting **Show Controls Palette** in the **Windows** menu. Once activated, the Controls Palette will appear whenever the front panel is brought to the foreground. It will disappear and be replaced by the Functions Palette when the block diagram is subsequently toggled into the foreground.

In **Controls>>Graph,** select a **Waveform Chart.** The Waveform Chart, one of three graphical modes available in LabVIEW, behaves like a laboratory strip chart, producing a real-time plot as each new data point is generated. In contrast, the **Waveform Graph** and **XY Graph** (which we will study later) display a full array of data that was produced at an earlier time.

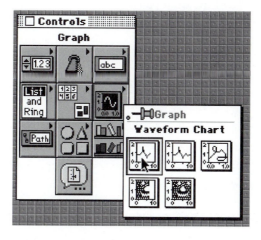

After you place the **Waveform Chart** on the front panel with a single mouse click, if you don't click the mouse further, a highlighted region will appear above the upper left of the plot. If you accidentally click the mouse so that the highlighted region disappears, don't fret; you'll learn how to make it reappear in a few minutes.

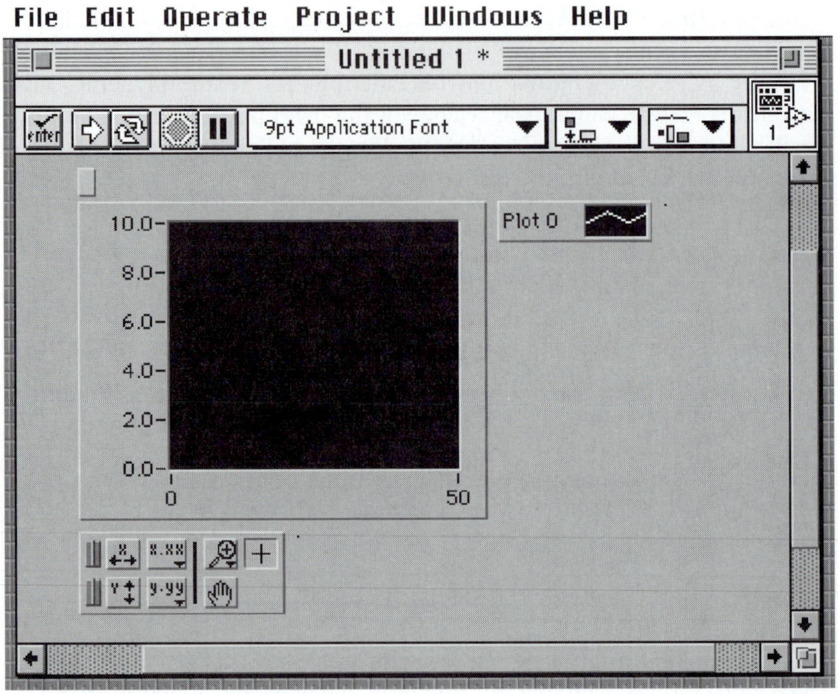

This highlighted region is the plot's *Label*. Using the keyboard, type a name for the plot such as **Waveform Chart**, then secure it by either clicking on the ⊞ button at the upper left end of the front panel, pressing *<Enter>* on the Numeric Keypad of your keyboard, or simply clicking the mouse cursor on an empty region of the front panel.

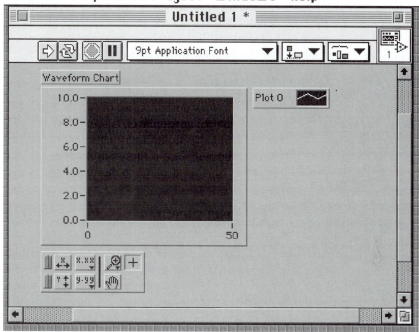

In addition to the chart region and the Label, the Waveform Chart includes a *Legend* and *Palette*.

As you will discover shortly, the Legend allows control over the plot style. Through it, one can choose such plot characteristics as whether data will be plotted as points or as interpolated lines, and the shape, if any, of the data points. The Palette, on the other hand, allows access to several useful functions that determine the scaling and labeling of the X and Y axes. Shortly, we will use the Palette to enable and disable an autoscaling feature of the Waveform Chart.

Let's explore how to reposition and resize the Waveform Chart. First, using the ↖ , highlight the entire Waveform Chart (including the chart region, Label, Legend, and Palette) with a marquee by placing the ↖ over the chart region and clicking. With the mouse button depressed, drag the highlighted object to the newly desired position. Once properly placed, release the mouse button and move the cursor to an empty spot on the block diagram. Then click the mouse to deselect the object. In addition, the Label, the Legend, and the Palette each can be moved independently. To demonstrate this feature, highlight just the Legend with a marquee by placing the ⏷ directly over this object and clicking. Then you will be able to drag this single object to a convenient location. Finally, to resize the chart region, position the ⏷ at one of the chart's corners. At the corner, the ⏷ will transform into a resizing handle ⏷. Click and drag this cursor to redefine the dimensions of your chart region. Experiment with this resizing for a few moments. You will find that you can resize both the actual chart region as well as its background frame.

Pop-Up Menu

You are about to enter a hidden world and ascend to a higher level as a LabVIEW programmer. You might wish to pause and reflect upon your life prior to the enlightenment that you are about to attain. Here is the deep secret: Almost all LabVIEW objects have their own associated *pop-up menu* that is hidden from the uninitiated. By gaining access to a pop-up menu, you, the programmer, are empowered to control the functioning of the associated object. How does one gain access to a pop-up menu? By performing the simple operation of "*popping up*" on the object. To pop up on an object, place the mouse cursor directly over it, then click the right mouse button (Windows) or click the mouse button while depressing the *<Cmd>* key on your keyboard (Macintosh). You can also pop up on an object by using the *Pop-Up Tool* ⏷ found in the Tools Palette.

Pop up on the chart region of the **Waveform Chart.** As a first example of controlling the features of this object using its pop-up menu, toggle the Label on and off by selecting **Label** from the **Show** palette. You may have toggled the Label off by an accidental extra mouse click when initially selecting the Waveform Chart from the Controls Palette. If so, use this opportunity to enter **Waveform Chart** for the Label. Also, for practice, you might try toggling on and off some other feature of the Waveform Chart such as its Palette or Legend. A word of warning about this pop-up menu: It is easy to inadvertently choose the first selection—**Change to Control**—which, when chosen, yields a "bad wire" on the block diagram. If this happens, the cure is to pop up and choose the first selection on the menu, **Change to Indicator.**

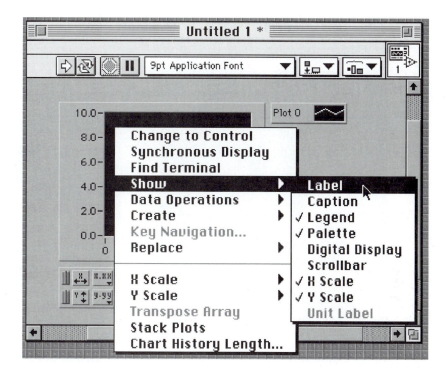

Scaling Plot Axes

A more important adjustment is the manner in which the axes should be scaled. In a moment you will be plotting sine-wave values on the Waveform Chart's Y-axis, so this axis must be prepared to chart data in the range of −1.0 to +1.0. Note that the default setting for the Waveform Chart's Y-axis data range is 0.0 to 10.0. To change the Y-axis data range from this default setting, select the Operating Tool 🖑 from the Tools Palette and use the 🖑 to highlight the upper limit of the Y-axis data range. Once highlighted, enter the newly desired value, which in our case is *+1.0*. Then in a similar manner, define the lower limit of the Y-axis data range to be *−1.0*. LabVIEW will then automatically redefine the intermediate Y-axis labeling.

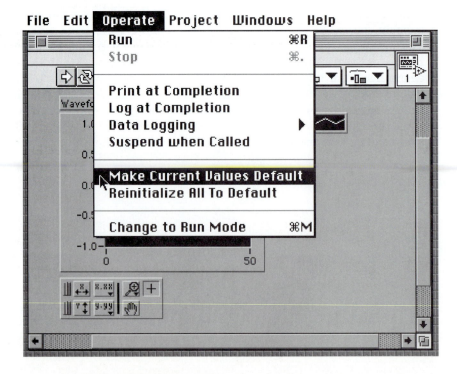

To save the redefined Y-axis labeling scheme, select **Make Current Values Default** from the **Operate** menu.

Autoscaling Feature

Alternately, the simplest method to scale the Y-axis appropriately is via activation of the Waveform Chart's autoscaling feature. There are two activation methods available to you. First, you can pop up on the chart region of the Waveform Chart and select **Autoscale Y** from the **Data Operations** palette.

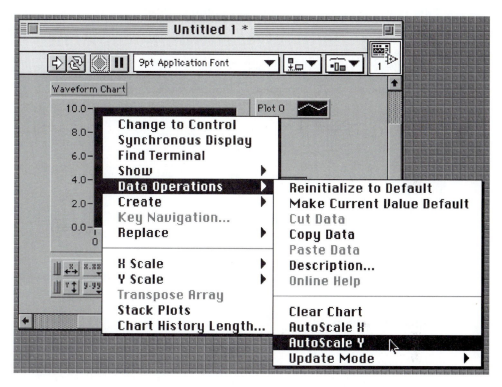

The second method is implemented on the Palette. Here, the top and bottom slider switches on the left side of the Palette enable and disable X-axis and Y-axis autoscaling, respectively. In the sequence of illustrations below, the Y-axis autoscaling slider is in the disable position (left), the LabVIEW-user clicks on the slider with the Operating Tool ✌(middle), which toggles the Y-axis autoscaling slider to the enable position (right). Try repeated clicks of the ✌, and watch the slider toggle back and forth between the enable and disable positions.

With one last editing step, you will be ready to run your first LabVIEW program. Use the keyboard shortcut *<Cmd><E>* (Macintosh) or *<Ctrl><E>* (Windows) to return to your block diagram. There you will find a Double Precision icon ▢DBL▢ that represents the data entrance to the Waveform Chart. It will be further identified by the

Waveform Chart label that you entered on the front panel. Using the ⬉ , place this icon near the **sin(x)** output of the Sine & Cosine icon, then use the ✦ to connect the **sin(x)** output to the ⬚DBL⬚ . Remember the Help Window can be activated with *<Cmd><H>* (Macintosh) or *<Ctrl><H>* (Windows) to aid in this wiring operation.

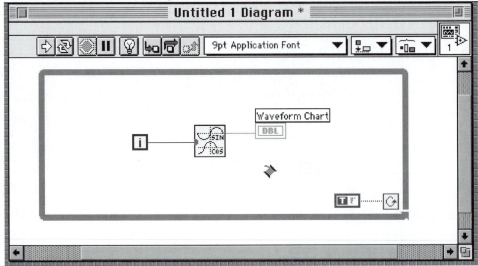

Program Execution

Now return to the front panel. You are ready to run your program, with the help of the *toolbar* shown here.

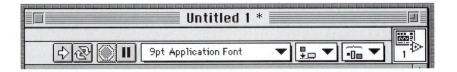

The leftmost button on the toolbar is the **Run** button ⬚. To start your program, simply click on the ⬚. As your program executes, you will observe a rather jagged-looking sine wave produced on your Waveform Chart in strip chart fashion.

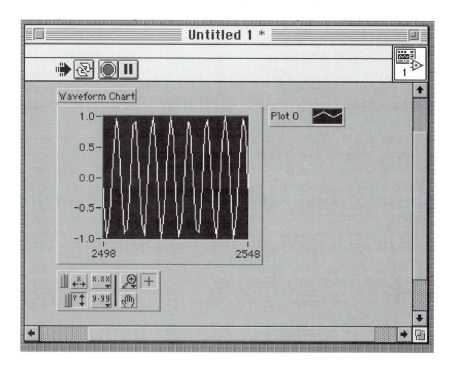

After watching the sine wave move across the chart for awhile, you may start to wonder about a few things. For example, we know that the block diagram is supplying Y-axis (sine-wave) values for the plot, but what does the X-axis represent? What determines the speed at which the sine wave appears to move? And, now that the program is running, how do we turn it off?

Let's answer the first question first. Concisely stated, the X-axis denotes the *count indices* of the plotted data. That is, the Waveform Chart keeps track of the number of data values it has received and associates a count index with each datum. For the ith data value, the plot displays an (X,Y) point, where X = i and Y = the actual data value. In its default setting, the Waveform Chart's X-axis values are the indices of the last 51 Y-axis (in our case, sine-wave) values supplied by the block diagram. You'll see how to modify this setting in a minute.

Now, addressing the second question, the speed at which the sine wave moves across the chart region is determined by the calculational time delay necessary to produce each new data point. So this speed simply reflects the time it takes for each iteration of your block diagram's While Loop. Because we are allowing this program to run freely, this time per iteration is determined by the speed of your computer's processor (something you can't change) and the settings of LabVIEW software switches (something you can change). To demonstrate this latter capability, use the Palette to toggle the **Autoscale Y** feature on and off. You may find that your program executes significantly faster with **Autoscale Y** off, because this feature requires some intensive calculational activity on the part of your processor. If your program runs just as fast when **Autoscale Y** is on as when it is off, count yourself fortunate, you have a fast computer (but you may be able to observe its speed limitations in the paragraph after next). Hopefully, this demonstration will drive home the point that a significant drop in performance sometimes must be accepted if handy and powerful programming features such as autoscaling are activated.

Finally, how do you turn the program off? On its block diagram, a TRUE Boolean value is constantly read by the While Loop's conditional terminal at the end of each loop iteration, so the program will run forever. Since we provided no way within the program to halt execution, your only recourse at the moment is to click on the **Abort Execution** button in the toolbar. Halt your program by clicking on the **Abort Execution** button.

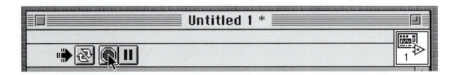

If you have a fast computer, try to observe your system's speed limitation in the following way. Pop up on the Waveform Chart and select **Clear Chart** from the **Data Operations** palette. Then, using the 🖑, enter *0* and *500* as the X-axis lower and upper limits, respectively. Now the Chart is set to plot the last 501 data points. Run your program and toggle **Autoscale Y** on and off using the Palette. See any slowing of your program with **Autoscale Y** on? If not, try overwhelming your computer with 1001 data-points calculations by repeating the procedure with initial X-axis lower and upper limits of *0* and *1000,* respectively. If that doesn't work, try 10,001 or 100,001 points. However, the Waveform Chart's default memory size is for 1024 data points. So, to plot more than 1024 points, you must modify the **Chart History Length...** setting in the Waveform Chart's pop-up menu (when the program is halted, and thus in its edit mode).

Now that you have used the **Abort Execution** button, let me mention that it is bad practice to do so. This button, at the moment it is activated, causes your computer to immediately cease the operation of your program. For the present program, such an action is probably no big concern. However, in more sophisticated programs, pressing **Abort Execution** at the wrong time could halt execution while reading data from a file or during communication with data acquisition boards in your computer's expansion slots. These situations lead to data corruption and other undesirable consequences. Thus it is always best to code a built-in stopping mechanism into your program.

Program Improvements

Based on these observations, let's improve your program in the following two ways: (1) provide a front-panel button that, when pressed, allows the program to complete its current While Loop iteration, and then halts the program, (2) provide a control over the speed at which the While Loop iterates.

Front Panel Switch.   LabVIEW provides a multitude of front-panel switches that can be used to bring your programs to a graceful stop. Halt your program so that you may edit its front panel. In **Controls>>Boolean,** select a switch such as the **Labelled Square Button.**

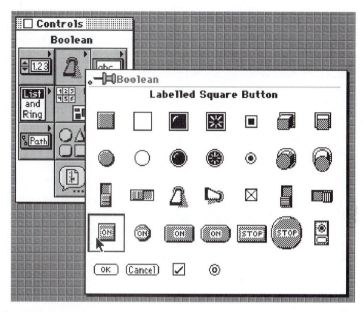

Using the , put this button in a convenient location. Notice that the default position for this switch is OFF, corresponding to a Boolean value of FALSE.

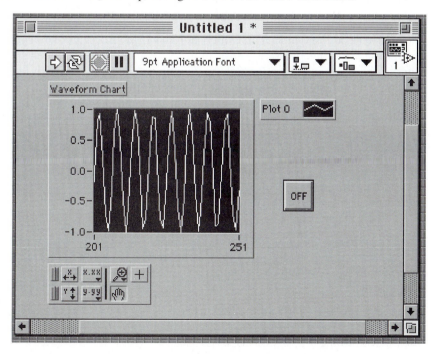

In controlling your program's execution, you will see momentarily that it is better for the switch default value to be TRUE. To select the button's ON state, click on it with the 🖑.

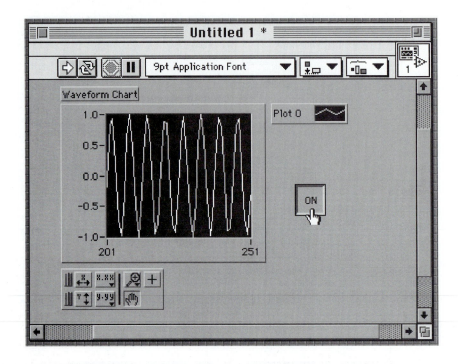

There are six modes in which a LabVIEW Boolean switch can behave in response to being pressed. These modes are listed in the switch's pop-up menu under the **Mechanical Action** palette. Pop up on the Square Switch and select **Latch When Pressed** from the **Mechanical Action** palette. In this mode, the switch changes to the opposite Boolean value when you click on it with the Operating Tool. It retains this new value until the program reads it once, at which point the switch returns to its default setting. You will find detailed explanations of each of the Mechanical Action modes in the LabVIEW User's Guide and in the Online Reference.

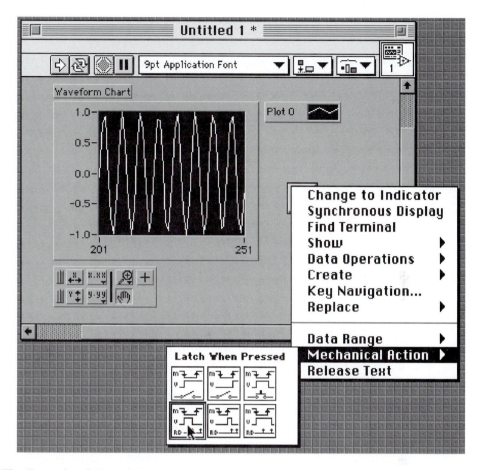

Finally, make **ON** and **Latch When Pressed** the button's default state by selecting **Make Current Values Default** from the **Operate** menu.

Now switch to the block diagram. There you will find the Boolean terminal TF that feeds the value of the front panel's Square Switch to the block diagram. Our plan is to connect TF to the While Loop's conditional terminal. To accomplish this, select the wire connecting the Boolean Constant to the conditional terminal by clicking on it with the ꓘ . Once it becomes highlighted with a marquee, erase this wire by pressing the ‹Delete› key on your keyboard.

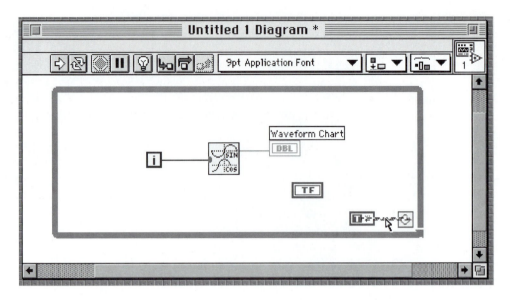

In a similar way, delete the **Boolean Constant** ▣. Now drag the Square Button's Boolean terminal ▣ near to the ▣ and wire these two objects together. Because you have made ▣ TRUE by default, the While Loop will continually iterate when the program is started. At some later time, almost assuredly during an intermediate portion of the While Loop cycle, the Square Button will be pressed OFF. Since the While Loop only checks the value of ▣ at the end of an iteration, the Loop will fully execute its calculations during that final iteration, before ceasing operation. Once the Square Button's OFF value is read, it will return to its default value of ON so the program is ready for the next time it is run.

Controlling Iteration Rate. While we are on the block diagram, let's also provide a method for controlling the rate at which the While Loop iterates. Through an icon called **Wait (ms),** LabVIEW provides a method of delaying subsequent program operations for a specified time period. We see from the Wait (ms)'s Help Window that, by specifying a numerical constant at the icon's input, the program will wait for that specified number of milliseconds.

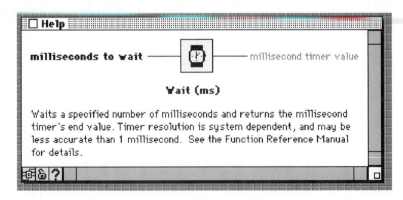

In **Functions>>Time & Dialog,** select the **Wait (ms)** icon and position it within the While Loop. In the Help Window above, the blue wire emanating from the **milliseconds to wait** input indicates that this terminal should be wired to an integer value. Select a **Numeric Constant** from **Functions>>Numeric.** When the **Numeric Constant** is first selected, its interior is highlighted and ready for you to enter a number. If you click the mouse, this highlighting will disappear. To restore the highlighting, use the Operating Tool. Enter the integer *300* into the **Numeric Constant** icon, then wire it to the input of **Wait (ms).**

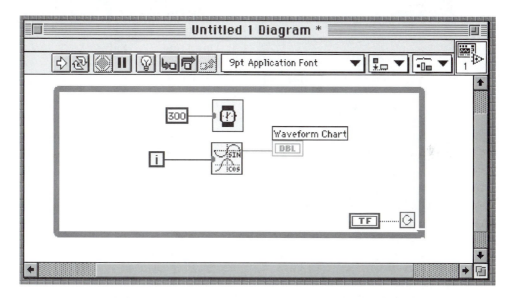

The previous block diagram causes one to ponder an obvious question. During each loop iteration, does the wait precede or succeed the sine-wave plot? The surprising answer is that LabVIEW effectively processes these two operations in parallel. That is, the wait and the sine-wave plot begin at the same time and the loop iteration concludes when both of these operations are complete. Thus, the time for an iteration is determined by whichever of the two is the slowest. Earlier we found that when only the sine-wave plotting was within the While Loop, the loop iterated much faster than three times a second. Thus, by including a delay of 300 milliseconds, this wait operation should provide the "rate-limiting step" that determines the time per iteration.

Now return to the front panel and run your new improved program by pressing the **Run** button ⇨. Because of the **Wait (ms)** icon, you will find the sine wave is now generated in a easy-to-observe manner. Use the front panel's Square Button to stop your program gracefully.

Improving Sine-Wave Resolution.   To see why your sine wave has a jagged appearance, pop up on the Waveform Chart's Legend. Under **Interpolation** you will discover that, by default, the Waveform Chart presents data in a "connect the dots" fashion. That is, a straight line is drawn from each data point to its neighbor and the entire collection of these lines represents the waveform. Also, the default **Point Style** is **None,** which means that no symbol is placed at each data point. However, you can make the actual data points visible by selecting, for example, a large **Solid Dot** for the **Point Style.**

While you're at it, you might like to spend a few minutes exploring the effect of selecting the various Point Style, Line Style, Interpolation, and Color options available.

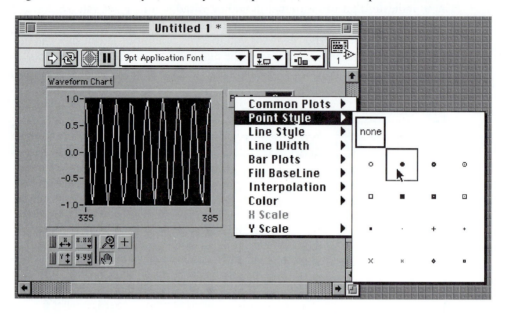

With the data points now visible, we see that each cycle of the sine wave is represented only by a few sampled locations. It is this sparse sampling that leads to the jagged waveform appearance when lines are interpolated between adjacent data points.

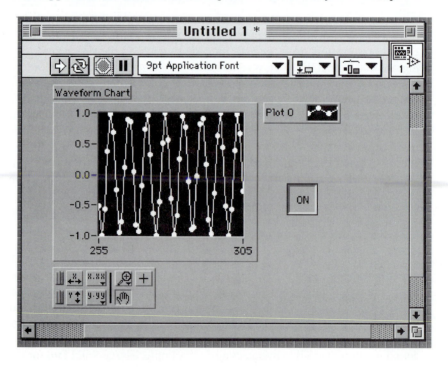

The reason for this paucity of data per cycle is, of course, due to our programming decision to take the sine function argument x to be the value of the While Loop's iteration terminal ⬚. Because ⬚ increments in steps of one and the sine function completes each new cycle every time x increases by $2\pi \approx 6$, only about six locations were sampled each sine-wave cycle.

To sample the sine function with higher resolution, let's take x to be one-fifth of ⬚, rather than simply ⬚. Then it will take five times as many iterations of the While Loop for x to increase by 2π, resulting in five times as many sampled points per cycle. To accomplish this feat, select the **Divide** icon from the **Functions>>Numeric.**

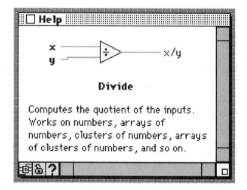

Then, also from **Functions>>Numeric,** obtain a **Numeric Constant.** When you first select the **Numeric Constant** (before the next mouse click), its interior is highlighted and ready to receive the numerical value of your choosing. Enter the number *5.0.* Once the mouse is clicked, the interior of the **Numeric Constant** can be highlighted again by using the Operating Tool.

Position these icons conveniently and wire them so that one-fifth of the ⬚ is input to the Sine & Cosine's argument **x** input. Note the gray coercion dot at the iteration terminal's connection to the **Divide.** This dot denotes that the integer from ⬚ is converted to a floating-point number. Thus the **Divide** icon performs a floating-point division operation.

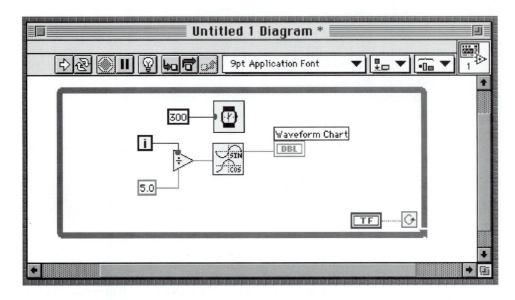

Run this final program and enjoy the beautifully realized sine wave being generated. While your program is running, imagine writing the mountains of code necessary to replicate this real-time sine-wave plot, complete with the attractive user-interface, using a text-based language such as C. I think you'll agree that the graphical-based LabVIEW programming language is simple, yet very powerful.

Data-Type Representations

There's one last perplexing detail to resolve before your block diagram is complete. You may have noticed that a coercion dot appears at the input of the **Wait (ms)** icon, indicating a numeric-format mismatch, although you seemingly have wired the requisite integer-formatted number to this input. This puzzle can be solved by consulting the Online Reference, which can be accessed by clicking on the 🔲 in the **Wait (ms)** Help Window.

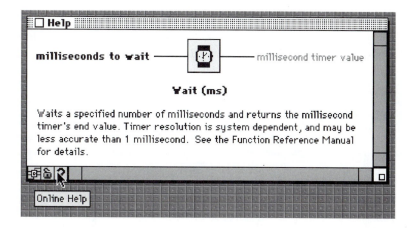

The Online Reference for **Wait (ms)** is shown here.

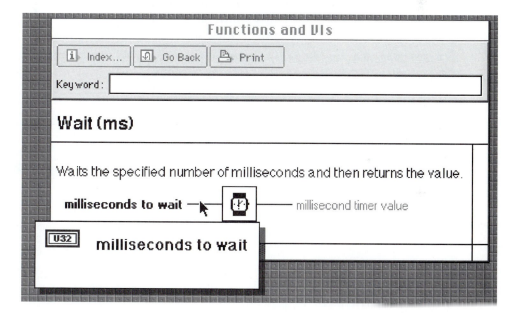

Clicking on **milliseconds to wait,** we find that this input is configured to accept an unsigned four-byte (or 32-bit) integer, a numeric format whose shorthand name is **U32.** This type of integer is always positive and can range in value from 0 to $(2^{32} - 1) = 4,294,967,295$. The unsigned integer-type is in contrast to the signed integer, which sacrifices one of its bits for use as a plus or minus sign. For example, the four-byte signed integer, a format called **I32,** can range from -2^{31} to $+(2^{31} - 1)$.

To discover the format of the integer on your block diagram, pop up on the 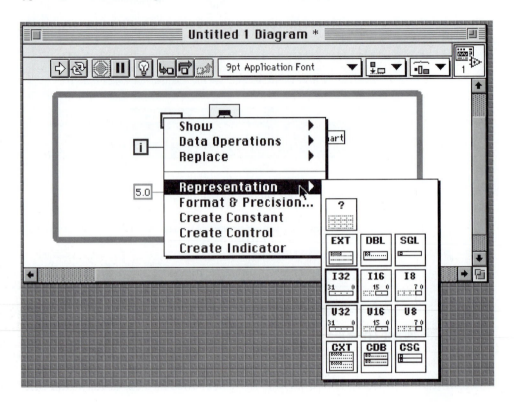, then select **Representation** from the pop-up menu. You will find that this integer is of type **I32,** the default integer format for a Numeric Constant.

Change the data-type of this number to an unsigned integer by selecting **U32** from the **Representation** palette. You will then find that the coercion dot disappears. Another mystery solved!

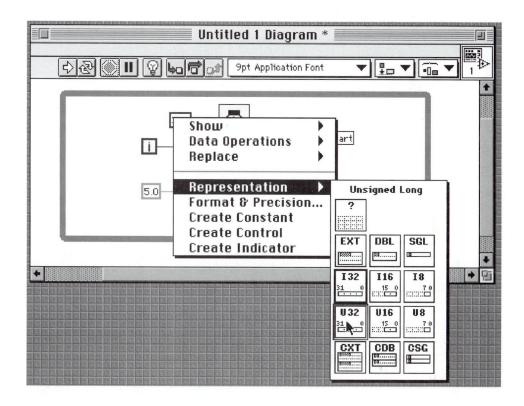

Automatic Creation Feature

Now that you understand some of the subtleties involved in wiring up constants to icon terminals, here's a time-saving shortcut I think you'll appreciate. First, delete the **Numeric Constant** 300 and its wire from your block diagram.

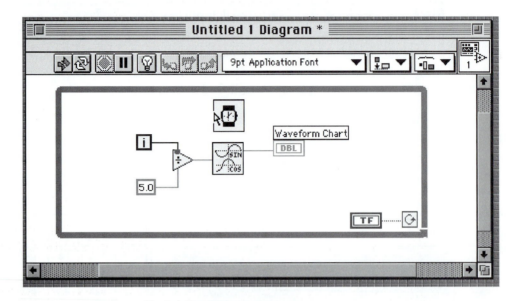

Now place the cursor over the **milliseconds to wait** input of **Wait (ms),** then pop up and select **Create Constant** from the menu that appears.

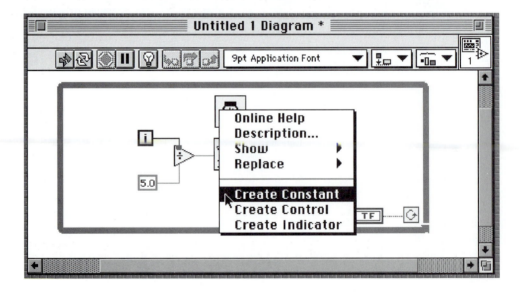

Like magic, an already-wired **Numeric Constant** of the correct data-type (in this case, **U32**) will appear, with its interior highlighted.

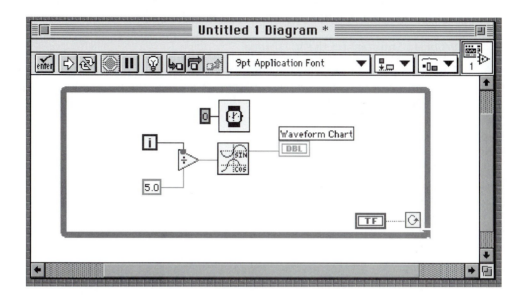

All you need to do is type in the desired integer value of *300* from the keyboard and you're finished.

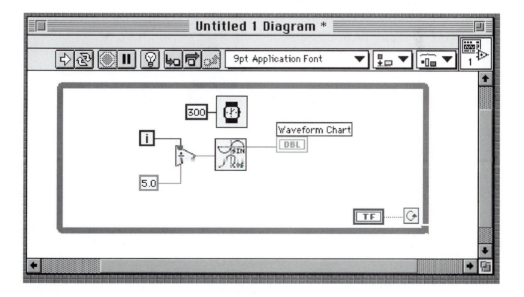

Automatic creation of "terminal-appropriate" objects is a universal feature of LabVIEW icons. Use of these time-saving "creation" options, available in the pop-up menus at the input and output terminals of all block diagram icons, will speed up your program development time.

Program Storage and the VI Library

LabVIEW programs simulate the function of laboratory instruments. As such, these programs are given the name *virtual instruments* (VIs). Your While Loop-based VI behaved as a sine-wave generator, so let's save this VI under a descriptive name such as **Sine Wave Generator-While Loop**. You can simply save your program in the LabVIEW folder (or directory), however, because you will be writing and saving many VIs in your study of LabVIEW, we'll use this opportunity to create a *VI Library*. This library will provide an easy-to-find location in which to store all of your programs. A VI Library contains compressed versions of your programs, so this method of storage has the added benefit of conserving disk space when compared to saving individual copies of each VI.

Here's how to create the VI Library and then store your VI in it. Under the **File** menu, select **Save As...**

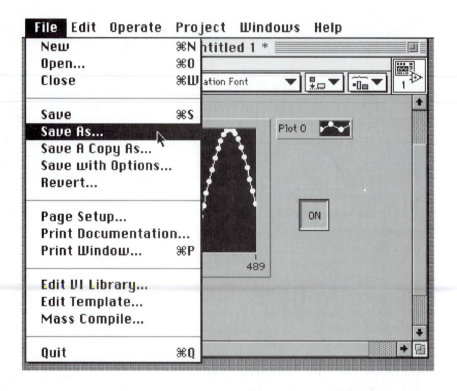

If you are using a Macintosh-based system, you will next see the window below. Select the **Use LLBs** button.

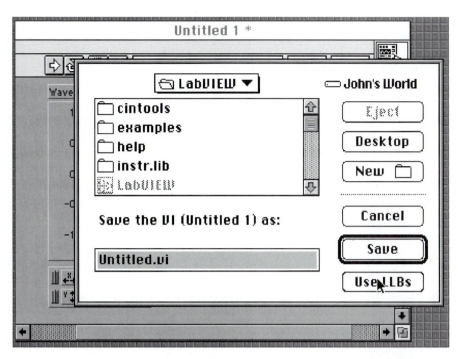

When the dialog box appears, create a VI Library by clicking **New...** (Macintosh) or **New VI Library** (Windows).

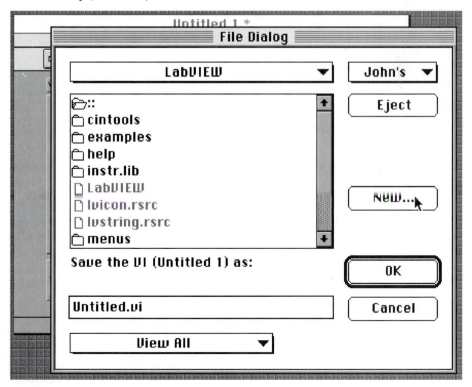

In the next dialog box, entitle the VI Library by typing **YourName**, then clicking on **VI Library**. LabVIEW will append the **.llb** extension when it names the new library. For Windows 3.1 users, since **YourName.llb** will be a file in your LabVIEW directory, **YourName** is limited to eight characters. For later versions of Windows and Macintosh users, you are free to use up to 255 and 31 characters, respectively.

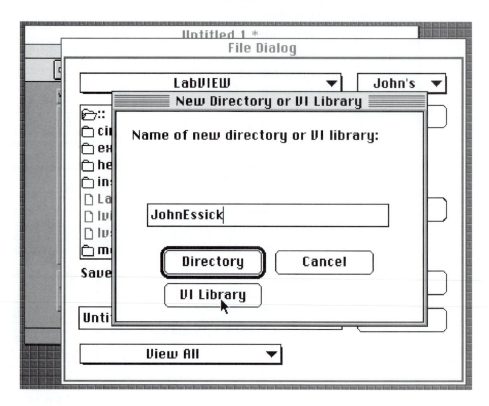

In the final dialog box, give your VI the name **Sine Wave Generator-While Loop**, regardless of your operating system (see below), in the **Save the VI as:** box, then store it under this name in the **YourName.llb** Library by clicking on the **OK** Button.

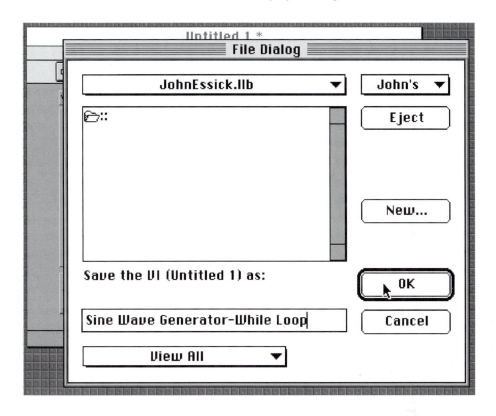

Since a VI Library (such as **YourName.llb**) is a file in your system, its name must abide by the restriction of your operating system. However, LabVIEW has reign over naming rules for the compressed VIs (such as **Sine Wave Generator-While Loop**) contained within a VI Library. LabVIEW stipulates that such VI names can contain up to 255 characters, including spaces and .vi extensions, and are not case sensitive.

CHAPTER 2

The For Loop And Waveform Graph

The LabVIEW programming language provides two possible loop structures to control repetitive operations in a program. In the previous chapter, we explored the While Loop, which executes the subdiagram within its borders until the Boolean value wired to its conditional terminal is FALSE. That is, the While Loop executes until a specified condition is no longer true. Now we will focus our attention on the *For Loop*, LabVIEW's other available loop structure. In contrast to the While Loop, which is controlled by the value of a specified condition, the For Loop simply executes the subdiagram within its borders a specified number of times. This loop structure is found in **Functions>>Structures** and appears on your block diagram as shown below.

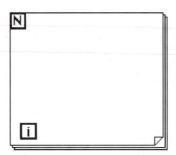

In this graphical structure, your "to-be-iterated" subprogram is written in the presently blank region within the borders. Once this subdiagram is written, the value of the *count terminal* ▣ determines the total number of times N that the For Loop will iterate. The value of the count terminal is set by wiring a **Numeric Constant** (located outside the loop) to the ▣. The For Loop's other internal icon is familiar from your work in the previous chapter. As in the While Loop, the iteration terminal ▣ contains the current number of completed loop iterations. The value of ▣ is *0* during the first loop iteration, *1* during the second, …, and $N - 1$ during the last iteration. Thus the For Loop is equivalent to the following text-based code:

```
For i = 0 to N - 1
    Execute Subprogram
```

In this chapter, you will write a Sine Wave Generator program based on a For Loop. In the process of writing this VI, you will explore LabVIEW's method of storing numerical data in the form of an array and learn to operate a *Waveform Graph*, the second in LabVIEW's triad of graphing options. Additionally, you will further hone your LabVIEW editing skills.

SINE WAVE GENERATOR USING A FOR LOOP AND WAVEFORM GRAPH

Create a fresh VI by selecting **New** under the **File** menu.

Here's a shortcut for activating the Controls Palette, if it isn't already visible: Pop up (right mouse-button click in Windows or *<Cmd>* click for Macintosh) on the blank front panel and a Controls Palette will appear. This palette is temporary, in that it will disappear as soon as you release the mouse button. However, you may secure the palette by placing the mouse cursor over the thumbtack located in the upper left corner of its window, then releasing the mouse button. This "tacked down" Controls Palette then can be placed in a convenient location by clicking and dragging its title bar. By the way, a similar thumbtacking procedure can be used to make any frequently used subpalette of the Controls or Functions Palette continually visible.

Place a **Waveform Graph** on the front panel from **Controls>>Graph**. Type in the label **Waveform Graph** and use the ⬧ to position the entire graph nicely on the front panel.

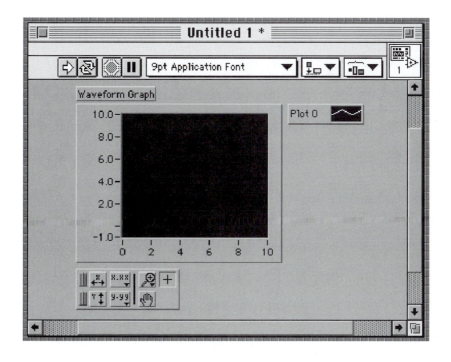

What is the **Waveform Graph** and how does it differ from the **Waveform Chart**? In our previous work, we saw that a **Waveform Chart** acts as a real-time data plotter. As each new data point is generated, this single numerical value is passed to the Waveform Chart's terminal ⬚DBL⬚ and then immediately displayed in the front-panel plot. The new data point is appended to the already existent data plot, so one can view the current value in the context of the previous values.

In contrast to the Chart's interactive method of plotting data, the **Waveform Graph** (as well as the yet unstudied **XY Graph**) accepts an entire block of data that has been generated previously. The Waveform Graph's terminal accepts this data block as a one-dimensional array of N numerical elements with the order of the elements indexed by integers in the range of 0 to $N-1$. For example, a 10-element array will have the following form.

Index	0	1	2	3	4	5	6	7	8	9
Array	1.20	1.30	1.40	1.50	1.60	1.70	1.80	1.90	2.00	2.10

The Waveform Graph then assumes that the numerical value of each element is the Y-value of a data point to be plotted. The X-value of each plotted point is taken to be the index of that point within the 1D array. That is, the X-value of the array's i^{th} element is i and the Y-value is the numerical value of the ith element. Thus the Waveform Graph implicitly assumes that data points are evenly spaced along the X-axis, a situation that is often realized in practice. For example, the time-varying nature of an analog signal is commonly represented by a set of discrete data points that were digitized during a sequence of equally spaced time intervals. You will find later that the XY Graph allows one to plot the more general case of an array of non-equally spaced data points (as well as multivalued functions such as circular shapes).

Switch to the block diagram. There you will find the Waveform Graph's terminal ⬚DBL⬚ along with its label. Note that the DBL (double-precision floating-point) data-type specifier contained within the terminal icon is enclosed in a square bracket. The square bracket indicates that this terminal should be wired to accept a numerical array, rather than just a single number as in the case of the Waveform Chart previously discussed.

Owned and Free Labels

Let me digress for a moment and discuss LabVIEW labels. The **Waveform Graph** label presently on your block diagram is what is known as an *owned label*. An owned label belongs to a particular object (in this case, the Waveform Graph's terminal) and moves with that object. To demonstrate this fact, use the Positioning Tool to highlight the terminal, then drag it to a new location on the block diagram. Note that the label accompanies the icon to its new location. Deselect the icon by clicking on an empty spot of the block diagram, then click on top of the label itself. A marquee will appear solely around the label, and you will be able to drag just the label to some newly desired configuration relative to its owning-icon. An owned label can be made visible or invisible using the **Show** command in the object's pop-up menu. Experiment a bit on the block diagram until you understand the functioning of an owned label.

An alternate form of LabVIEW annotation is called the *free label*. Free labels are not attached to any object and can be created, moved, or disposed of independently. They can be used as explanatory comments on your front panels and block diagrams. To create a free label, select the *Labeling Tool* **A** from the Tools Palette and click on the desired location for the label. A small bordered box appears, ready to accept text input. After you type the text message, press *<Enter>* on the Numeric Keypad, and the free label is complete. Try producing a free label on your block diagram. After you succeed in creating a free label, click on it with the Positioning Tool and delete it.

Now select a **For Loop** from **Functions>>Structures** and add it to your block diagram (the Functions Palette can be activated by popping up on a blank region of the block diagram). If you are dissatisfied with your initial choice of dimensions, you can place the Positioning Tool at the corner of the For Loop where it will morph into the resizing handle and allow you to reshape the loop borders.

Our programming strategy is this: Use the For Loop to generate a one-dimensional array of data points that represent a few cycles of the sine function, then pass this array to the Waveform Graph's terminal for plotting on the front panel. Since we desire data to be passed to the terminal only after the For Loop has completed its final iteration, we must place the terminal icon in a region outside the Loop's boundary. Use the Positioning Tool to create the following block diagram.

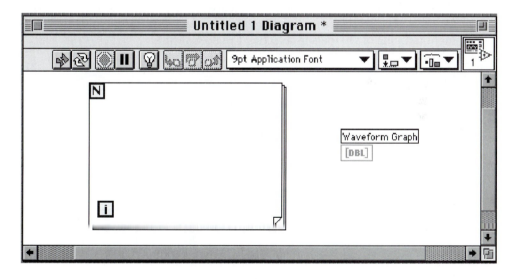

Now let's write the code for generating the sine-wave values within the For Loop. To achieve this goal, obtain a **Sine & Cosine** icon from **Functions>> Numeric>>Trigonometric** and place it inside the For Loop.

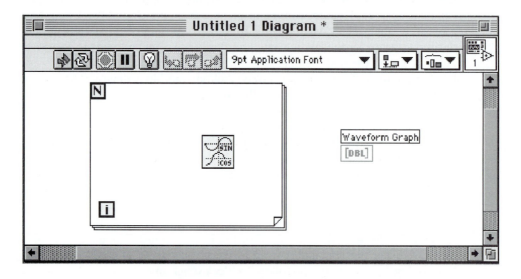

Now use **Divide**, from **Functions>>Numeric**, to write the sine-wave generation code shown below. Try creating the **Numeric Constant** 5.00 by popping up on the lower input terminal of **Divide** and selecting **Create Constant**.

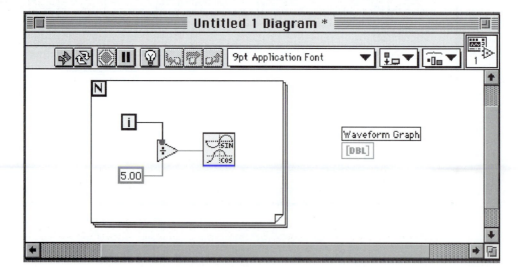

Ready for another editing shortcut? When coding a block diagram, you will find that the most commonly used tools are the and . Rather than having to go to the Tools Palette each time to switch back and forth between these tools, simply press the *<Spacebar>* on your keyboard. With each depression of the *<Spacebar>* key, the mouse cursor will toggle back and forth between the Positioning and Wiring Tool. If you require some other feature in the Tools Palette, press your keyboard's *<Tab>* key. As you press *<Tab>* repeatedly, the mouse cursor will cycle through all the possible tools in the Tools Palette.

Now we need to tell the For Loop the number of iterations N we would like it to perform. We know from our experience in the last chapter that about 30 loop iterations will be required to generate one cycle of the sine function. Let's set N=100, so that we generate (a little more than) three sine-wave cycles. To accomplish this setting, wire a **Numeric Constant** (defined to be *100*) to the . You already know two methods for acquiring the required Numeric Constant icon—the "hard" way by obtaining it directly from **Functions>>Numeric** and the "easy" way by simply popping up on the and selecting **Create Constant**. Be good to yourself—use the "easy" way.

Cloning Block-Diagram Icons

Here's one final editing option that you can add to your bag of tricks. It's called *cloning* and can be used in the present situation to produce the required icon because an equivalent icon already exists on your block diagram. To clone the existing **Numeric Constant** 5.00 on your block diagram, place the Positioning Tool over this icon.

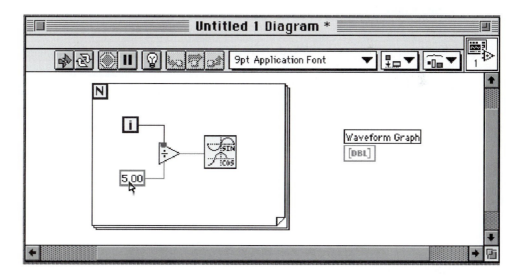

Now click the mouse button while depressing the *<Option>* (Macintosh) or *<Ctrl>* (Windows) key on your keyboard. Then, by moving the mouse while still holding down the button, you will drag a copy of the icon (displayed as a dotted line) with you, while the original stays in place.

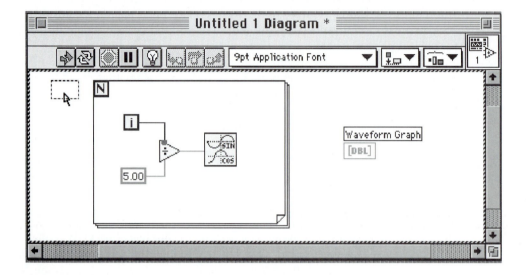

Once you arrive at the newly desired location, release the mouse button and the cloned icon will appear.

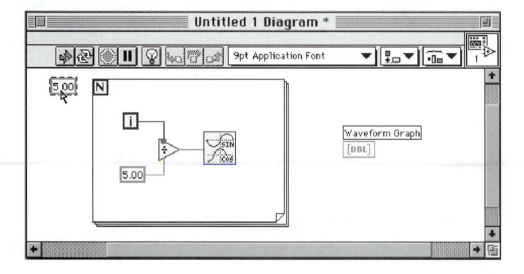

Use the Operating Tool to change the value of this new **Numeric Constant** to *100* and then wire it to the .

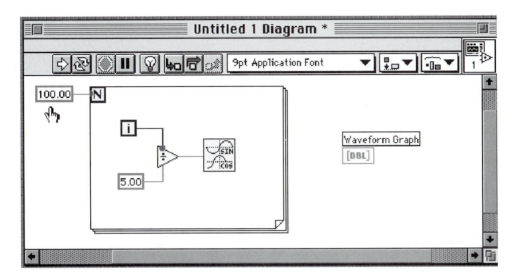

The For Loop's count terminal is configured to accept the numeric format **I32** (four-byte, signed integer). The coercion dot, indicating a data-type mismatch, can then be erased by popping up on the **Numeric Constant** and selecting **I32** in the **Representation** palette.

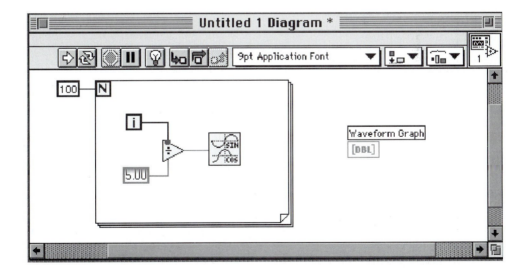

Finally, we need to store the sequence of sine-function values produced by the For Loop in a 100-element array so that we can pass this array to the Waveform Graph's terminal for plotting. Does it sound like there's some hard programming ahead? Surprisingly, there's not! Array storage of the sequence of values generated by the multiple iterations of a loop structure is such a commonly required operation that LabVIEW provides it as a built-in optional feature of both the For and While Loops. Perhaps it's easiest if we first implement this feature and then decipher what we've done.

Auto-Indexing Feature

Using the ✎, wire the **sin(x)** output of the Sine & Cosine icon to the Waveform Graph's terminal. You're finished!

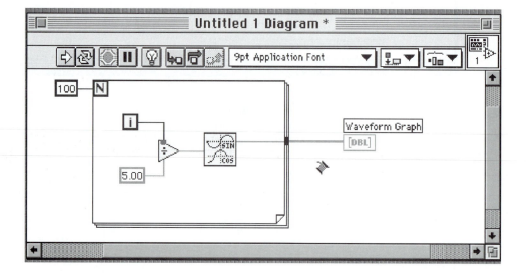

Here, you've accomplished our goal by implementing the auto-indexing capability of the For Loop. With this feature, as the For Loop creates a new value with each loop iteration, this new value is indexed and appended to an array at the loop boundary. After the loop completes its final iteration, the array is passed out of the loop to the Graph's terminal. Notice that, in the region within the For Loop, the wire emanating from the Sine & Cosine output is thin, denoting the fact that this icon is producing just a single numerical value with each loop iteration. However, when this wire passes through the black rectangle at the loop's boundary (called a *tunnel*), the wire becomes much thicker. This thick wire is LabVIEW's way of denoting a one-dimensional array of data values.

Let's explore this auto-indexing feature a little further. Place the Positioning Tool over the tunnel at the For Loop's boundary.

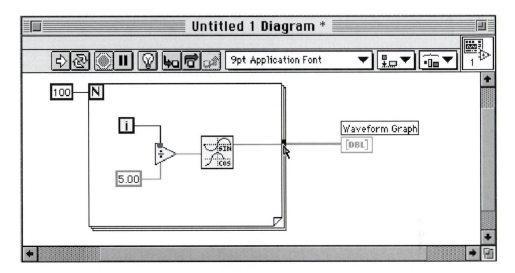

Then pop up on this tunnel to reveal its options menu.

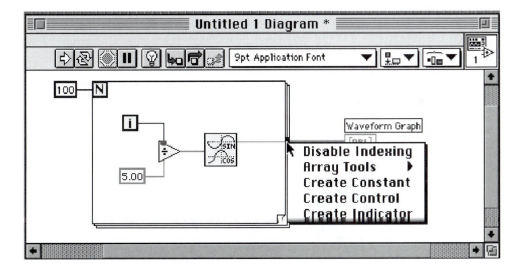

This menu offers you the ability of toggling the loop's auto-indexing feature on and off. A For Loop's default setting is **Enable Indexing**, so the menu now offers the option of toggling this feature off with the selection **Disable Indexing**. Toggle the auto-indexing off by selecting **Disable Indexing**.

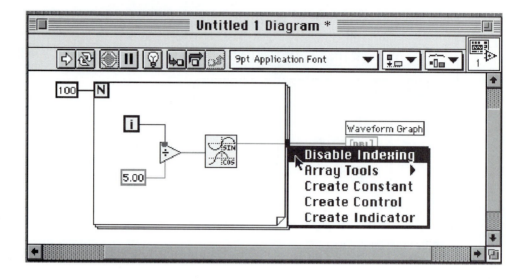

Now, rather than accumulating all the loop's iterated values into an array, the loop will only pass out a single value, the value of the sine function determined during the last loop iteration. As shown here, with the auto-indexing disabled, a bad wire now appears connected to the Waveform Graph's terminal because this icon expects to receive an entire array, not just a single numerical value.

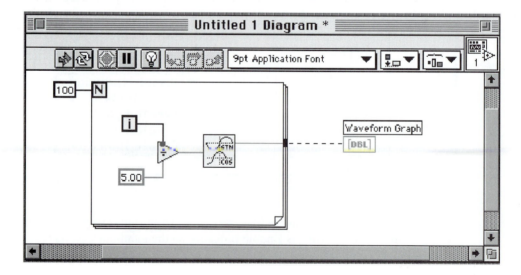

To restore the proper programming, pop up on the tunnel and toggle the auto-indexing back on by selecting **Enable Indexing**.

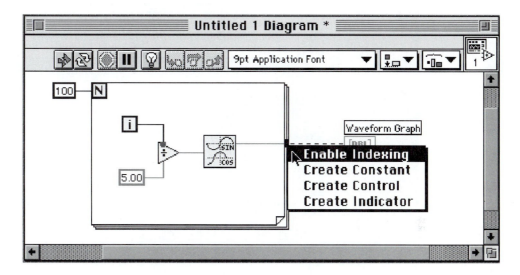

To see if your program works, return to the front panel and use the Operating Tool to redefine the range of each plot axis. You will be plotting a sine wave discretely sampled at 100 points. Thus the Y-axis values should range from −1.0 to +1.0. The X-axis value of each sine-wave value is taken to be its index within the one-dimensional array of data points. Because LabVIEW ascribes an index value of 0 to the first array element, the index of the last datum is 99. Thus the X-axis values range from 0 to 99. Perhaps it is more pleasing to choose 100 as the X-axis upper limit. Use the Operating Tool to define the upper and lower limits on each axis. LabVIEW will automatically fill in the intermediate label values. To save these settings, select **Make Current Values Default** from the **Operate** menu.

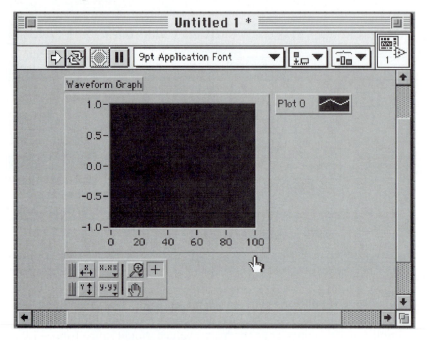

Alternately, you can enable the autoscaling on the two axes either through the Graph's pop-up menu or the Graph's Palette.

Now run your program and hopefully it will produce a plot as shown.

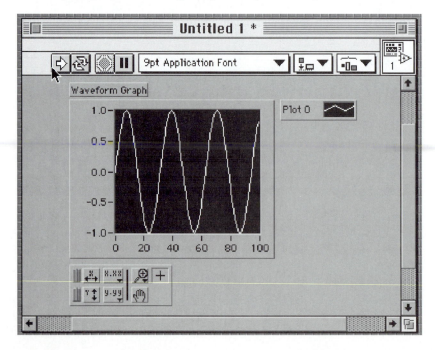

X-Axis Calibration of the Waveform Graph

Pretty nice program, eh? There is one thing, however, we should feel the need to improve—the X-axis labeling. As is, this labeling reflects the integer indexing derived from the array storage of your data. It would be much more desirable to have this axis calibrated in some angular measurement appropriate to the argument of the sine function being plotted on the Y-axis. Remembering that the program takes the sine function argument for a particular data point to be the array index value *i* divided by 5, we see that the X-axis can be calibrated (in radians) by simply multiplying each label by 0.2. Alternately, we can re-express this idea by saying that, by default, the Waveform Graph assumes the X-axis spacing between adjacent data points Δx is 1. Calibration is then accomplished by defining Δx to be 0.2.

Such calibration by a multiplicative constant is a common need when using the Waveform Graph and thus LabVIEW provides a mechanism for this procedure. This calibration mechanism involves a new LabVIEW process: bundling together several objects into a *cluster*. The resulting cluster in not a mathematical object, but rather a LabVIEW convenience. By grouping a number of objects together into a single cluster, one is able to transport a large quantity of data across a block diagram using a single cluster wire. Thus clustering greatly simplifies the appearance and readability of your block diagrams.

Return to the block diagram, place the mouse cursor over the Waveform Graph's terminal, and activate its Help Window.

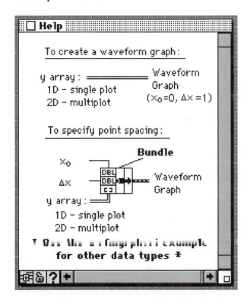

This Help Window, which is divided into a top and bottom portion, appears above. In the top portion, we find that a Waveform Graph with default data-point spacing $\Delta x = 1$ is created by simply connecting an array wire to the Graph terminal. This is the manner in which we've used the Waveform Graph thus far. We produced a single plot by feeding a one-dimensional data array to the Graph's terminal. Note that two (or more) waveforms can be simultaneously displayed on a Graph by connecting a 2D array to the terminal. The bottom portion of the Help Window indicates that, when given an appropriate cluster, the Waveform Graph produces a plot on the front panel with its X-axis properly calibrated. The appropriate cluster is formed by bundling together the following sequence of objects: a value for the X-axis origin x_0, a value for the X-axis spacing Δx and the array of numerical data to be used as the Y-axis values of the waveform's points.

Here's how to include the appropriate cluster in your code. First, remove the array wire connected to the Graph terminal. Using the Positioning Tool, click on this wire to produce a marquee.

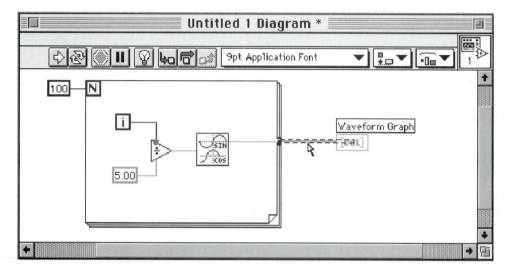

Then press your keyboard's <*Delete*> key to remove this wire.

Select **Bundle** from **Functions>>Cluster**. The **Bundle** icon is used to collect several input objects into a cluster.

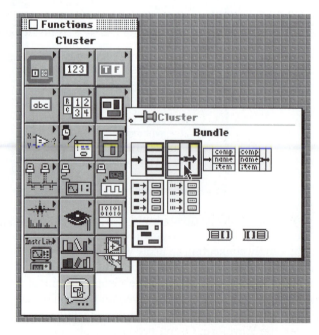

Then place the **Bundle** icon on your block diagram. You will find that its leftmost section consists of two inputs by default. We require three inputs, so you must resize the icon.

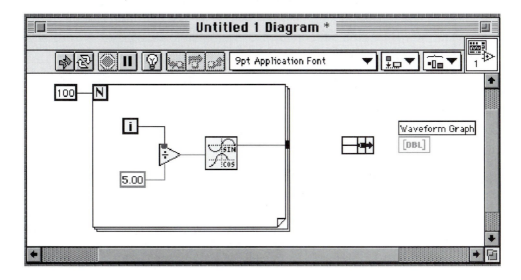

To resize the **Bundle** icon so that it has three inputs, place the Positioning Tool at one of the icon's corners, so that the tool morphs into a resizing handle.

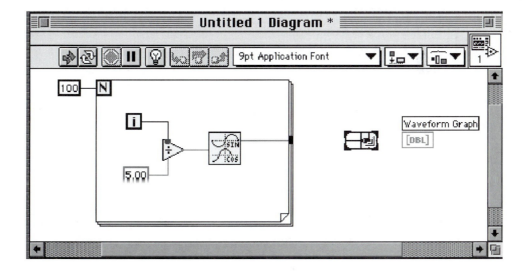

Then drag the handle until an additional input appears.

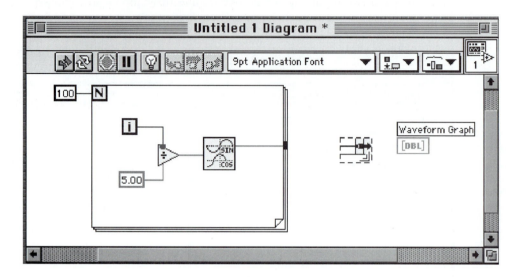

Secure this choice of inputs by releasing the mouse button. Use the Positioning Tool to position the Bundle and Graph terminal icon as shown.

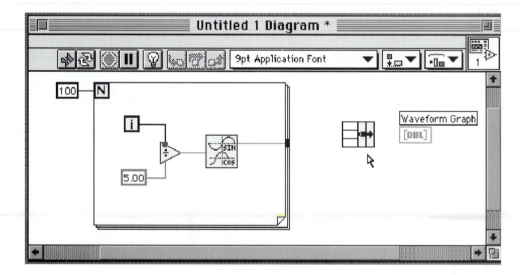

Wire the For Loop's tunnel to the Bundle's bottom input. Once this connection is made, square brackets will appear within this input, the graphical specifier for an array data-type.

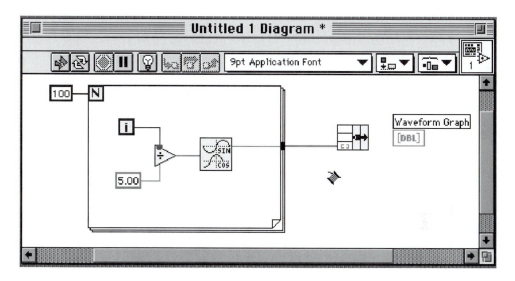

Now wire a **Numeric Constant** of *0.0* and *0.2* to the Bundle's top and middle inputs to define x_o and Δx, respectively. Note that unbracketed DBLs appear within these inputs, indicating that each consists of a solitary (i.e., non-array) floating-point number.

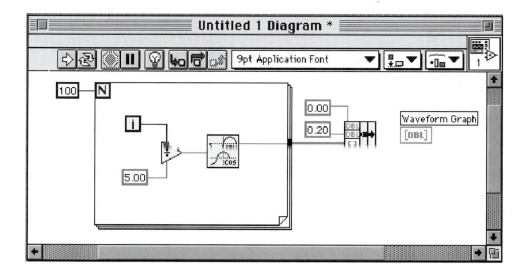

The Bundle icon receives these three inputs on its left side, bundles them together, and outputs the resulting cluster at its rightmost section. Wire this cluster output to the Waveform Graph's terminal.

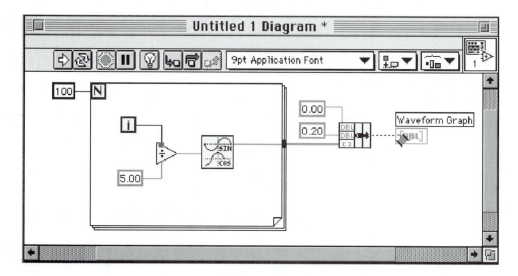

You will find that as soon as this connection is made, the terminal's icon morphs from an array specifier [DBL] into a cluster specifier [⊞]. Also the wire connecting the Bundle output to the Graph's terminal has a braided appearance, denoting cluster data.

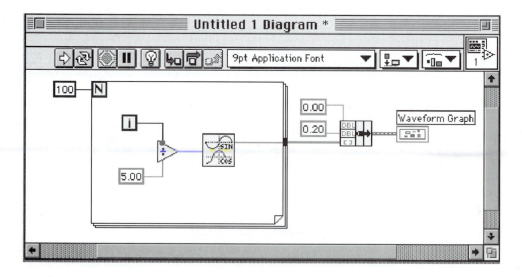

Now return to the front panel and run your final program. Note that the X-axis is calibrated in radians. You may want to activate autoscaling on the X-axis.

Using **Save** in the **File** menu, save your VI in the **YourName.llb** Library under the name **Sine Wave Generator-For Loop**.

SINE WAVE GENERATOR USING A WHILE LOOP AND WAVEFORM GRAPH

Let's try accomplishing the above goal of plotting a sine wave on a **Waveform Graph**, but this time using a While Loop rather than a For Loop to generate the data array. Select **New** from the **File** menu, then place a **Waveform Graph** (from **Controls>> Graph**) on the fresh front panel. Enable **Autoscale X** and **Autoscale Y** using the Palette.

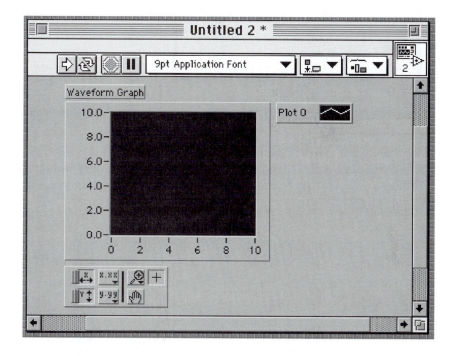

Switch to the block diagram and write the following code.

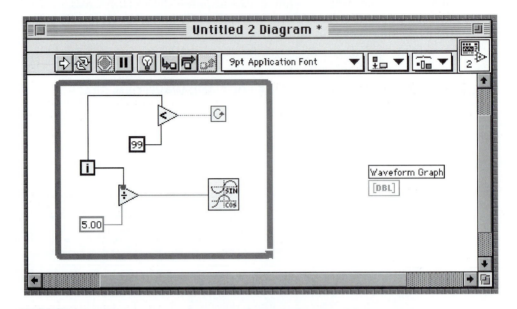

The **Less?** icon is found in **Functions>>Comparison**. Its function is described by the Help Window shown below. Remember, you can always access assistance from such Help Windows while wiring by toggling on **Show Help** in the **Help** menu or, more simply, through the keyboard shortcut: *<Cmd><H>* (Macintosh) or *<Ctrl><H>* (Windows).

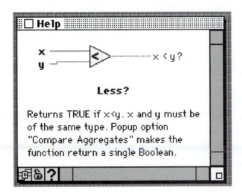

The above While Loop will generate 100 values of the sine function before ceasing execution. Using your knowledge of the detailed operation of a While Loop, can you explain why this is so? In particular, explain why the proper value for the **Numeric Constant** in the comparison statement is *99* (and not *100*).

You now need to store the 100 sine-function values in a one-dimensional array and pass this block of data to the Waveform Graph's terminal for plotting on the front panel. Making use of the auto-indexing feature of the loop structure, you need simply to wire the **sin(x)** output of the Sine & Cosine icon to the Graph terminal. Perform this wiring as shown here.

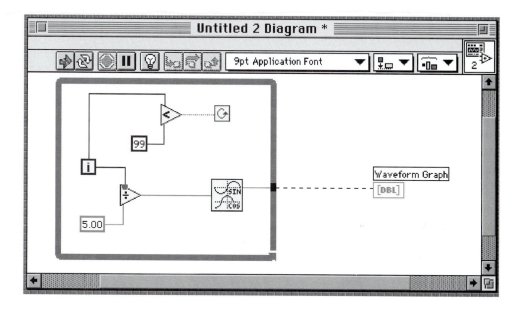

You will note that this seemingly simple connection produces a bad wire. To find out why, pop up on the tunnel at the While Loop's border.

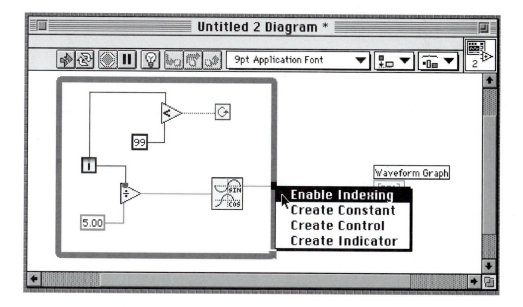

In the tunnel's pop-up menu, you will be presented the option of enabling the loop's auto-indexing feature. Thus we see that, in contrast to a For Loop, the auto-indexing feature of a While Loop is toggled off by default. Turn the auto-indexing on by selecting **Enable Indexing**. Also include a **Bundle** icon to scale the X-axis in radians, as shown here.

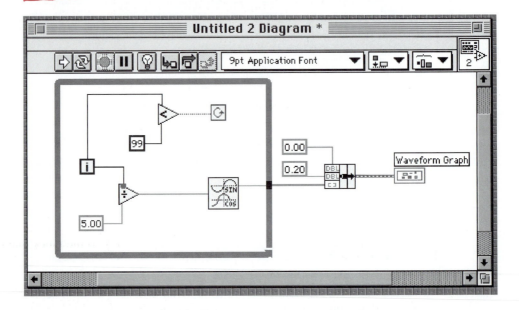

Your program is now complete. Return to the front panel and run it. You might try using the keyboard shortcut for the **Run** command: *<Cmd><R>* (Macintosh) or *<Ctrl><R>* (Windows).

Array Indicators and the Probe

Are you unconvinced that there are actually 100 elements in the array you produced? You can easily check this fact by inspecting the array with an *Array Indicator*. An Array Indictor can be constructed as follows. Select an **Array** shell from **Controls>> Array & Cluster**. As described in the next paragraph, the **Array** shell can be made to operate in either of two modes: to input an array from the front panel to the block diagram or to output an array from the block diagram to the front panel.

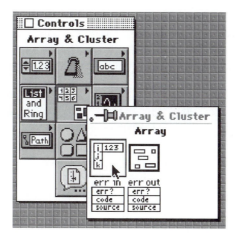

Place the **Array** shell on your front panel, then enter the label **Array**. You will find that the shell consists of an *index display* on the left and a large blank region on the right. By filling the blank region with the appropriate object, one may choose whether the **Array** shell will function as an input or output device. In the present situation, we wish to output an array to the front panel, so we will occupy this blank region with a **Digital Indicator**. If instead we wanted to input an array to the block diagram, we would occupy this region with a **Digital Control**.

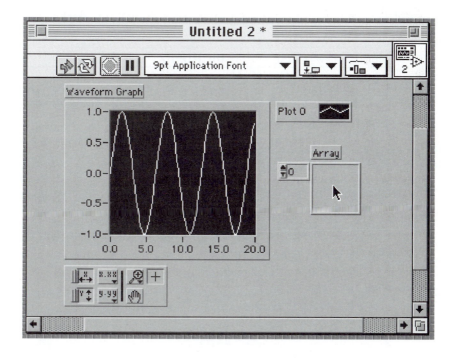

To place a **Digital Indicator** within the **Array** shell, do the following. Select a **Digital Indicator** from **Controls>>Numeric**.

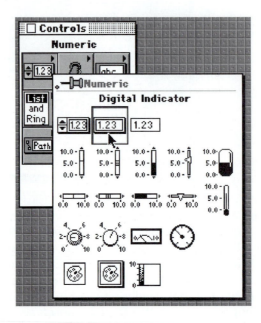

Then place the cursor in the large blank region of the **Array** shell.

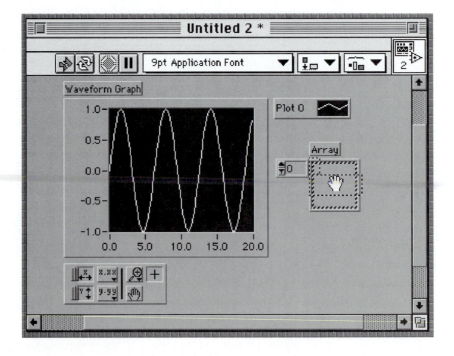

When you click the mouse button, the Array Indicator will be complete as shown below.

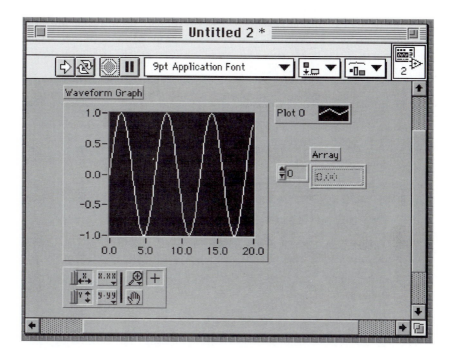

Now go to the block diagram and place the Array Indicator's terminal in a convenient location. In this program, the While Loop creates a 1D array of data. Using the Wiring Tool, connect any point on the thick wire representing this 1D array to the Array Indicator's terminal.

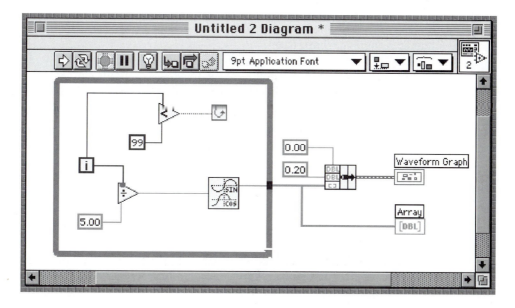

Return to the front panel and run your program. After the VI completes execution, you can inspect the array generated via the Array Indicator. Using the Operating Tool, look at various elements in the array. To scan the array, you may increment or decrement the index number in steps of 1 by repeatedly clicking on the index display's up/down arrows. Alternately, you can use the Operating Tool to highlight the index display, enter the newly desired value from the keyboard, and then press <*Enter*>, <*Return*>, or click on the ☑.

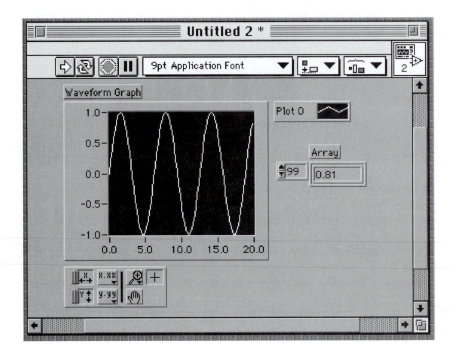

Through such exploration, you should find that your array consists of 100 elements, with indices running from 0 to 99.

LabVIEW's *Probe* provides another method for checking the contents of your array. This handy feature allows you to examine the contents on any wire on your block diagram. To inspect the elements of the sine-wave array using the Probe, place the mouse cursor on the wire carrying this data.

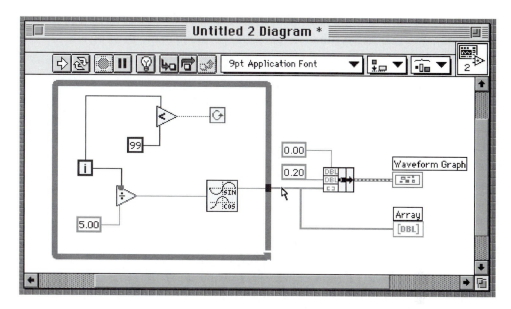

Then pop up on this wire and select **Probe**.

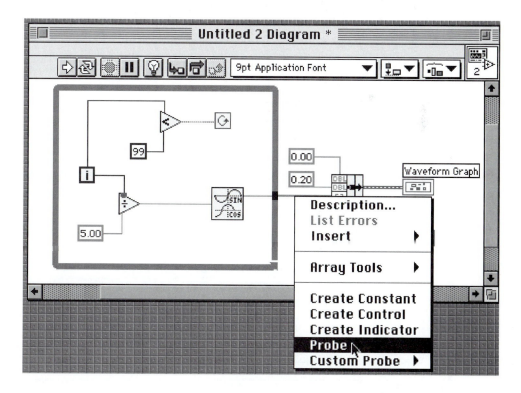

A Probe Window will now appear. Probe Windows can also be activated by clicking on the wire of interest with the *Probe Tool* ⊕ found in the Tools Palette.

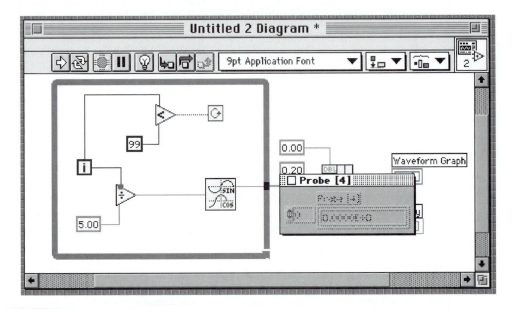

Run your program, then use the Probe to inspect the elements of the resulting array. When you're done with the Probe, close its window and it will disappear. Probe Windows are an invaluable tool for debugging LabVIEW programs that execute, but are producing questionable or unexpected results.

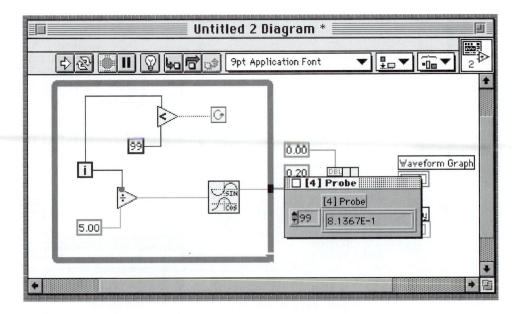

Finally, you might try producing an Array Indicator the painless way. For example, pop up on the tunnel at which the sine-wave array outputs from the While Loop.

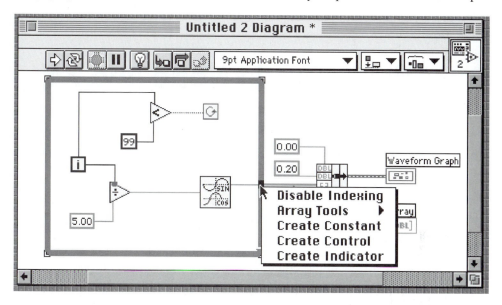

Then select **Create Indicator** from the pop-up menu.

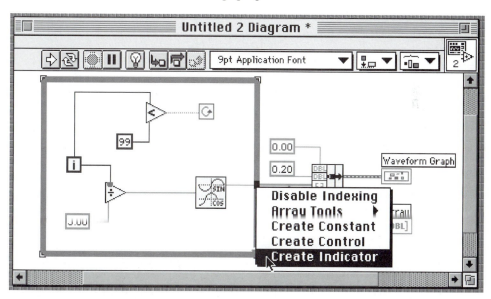

Now check the front panel. You will find an automatically produced Array Indicator there. Run your program and discover if it works as expected. You might also try creating a front-panel indicator by popping up on the Waveform Graph's cluster terminal.

Using **Save** in the **File** menu, save this VI in the **YourName.llb** Library under the name **Sine Wave Generator-While Loop 2**.

CHAPTER 3

The Formula Node and XY Graph

In previous chapters, we have seen that the graphical nature of the LabVIEW programming language greatly simplifies the coding of desired computer operations. However, in the case of encoding algebraic equations, graphical programming is much more cumbersome than text-based languages. For this reason, LabVIEW offers the *Formula Node*, a resizable box that you can use to enter text-based algebraic formulas directly on the block diagram. To demonstrate the utility of the Formula Node, consider the relatively simple equation $Y = 3X^2 + 2X + 1$. If you code this equation using LabVIEW arithmetic icons from **Functions>>Numeric**, the subdiagram will appear as below.

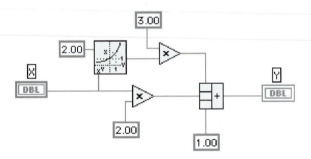

Alternately, you can program the same equation using a Formula Node, which is found in **Functions>>Structures**. The resulting subdiagram is shown in the following illustration. Note that, within the Formula Node, an equation is terminated by a semicolon (;).

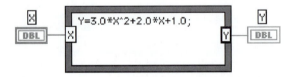

As you can see, the Formula Node is much easier to write and to read. All manner of mathematical operations are possible within a Formula Node, including trigonometric and logarithmic functions as well as Boolean logic, comparisons, and conditional branching. The text-based functions available within a Formula Node are listed in its Help Window, which is reproduced here.

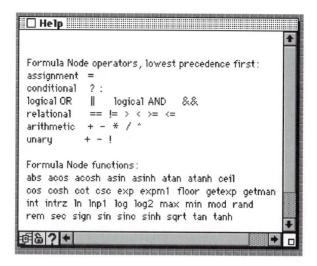

Also, the capacity of a particular Formula Node is not limited to a single equation. This structure can be resized to contain as many equations as you desire. You will find that when coding a complicated formula or when evaluating a group of equations, the Formula Node is the way to go.

In this chapter, you will rewrite your Sine Wave Generator using a Formula Node. You will then use this subdiagram to supply data to an *XY Graph*, the most general of LabVIEW's three graphing options. This new program will become the basis for **Sine Wave**, a VI you will use to simulate discretely sampled data in your future work with File I/O and Fast Fourier Transform-based programs. Finally, you will learn how to create your own custom-made icon that can then be used as a *subVI* in larger "calling" programs.

SINE WAVE GENERATOR USING A FORMULA NODE

On a new front panel, place a **Waveform Graph** and label it **Waveform Graph**. Using its pop up menu or Palette, activate the **Autoscale X** and **Autoscale Y** features of the Graph, if desired. You also can choose to hide the Graph's Palette and Legend using **Show** in the pop-up menu.

Place a **Digital Control** (found in **Controls>>Numeric**) on your front panel. You will use this control to tell the program how many data points you'd like it to generate, so give it the label **Number of Points**. Finally, place an Array Indicator (select **Array** shell from **Controls>>Array & Cluster**, then place a **Digital Indicator** from **Controls>> Numeric** within it) on the front panel and label it **Y Array**. If you aren't pleased with the positioning of a particular owned label, click directly over it. It will become highlighted by a marquee and you will be able to move it independently, that is, without moving its associated object.

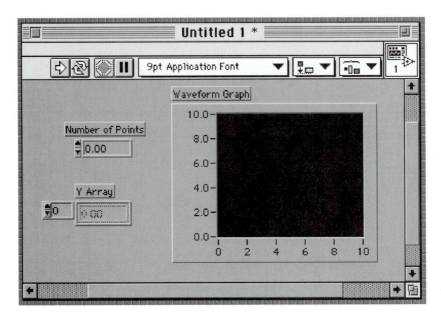

Switch to the block diagram and place a **For Loop** on it. Wire the **Number of Points** terminal to the , so the For Loop's total iteration count is controllable from the front panel.

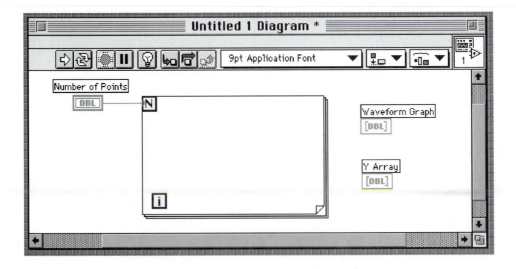

The For Loop's count terminal is configured to accept a four-byte signed integer input. To avoid a coercion dot at the , pop up on the **Number of Points** terminal and change **Representation** from its default format of **DBL** to **I32**.

In the previous chapter, we generated a sine wave with a subdiagram composed of LabVIEW's arithmetic icons. Now let's write the equivalent code but this time using a Formula Node. Select a **Formula Node** from **Functions>>Structures** and position it within the borders of the For Loop.

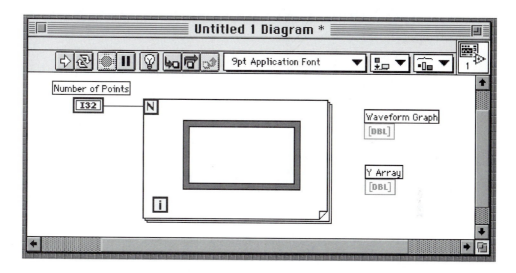

Place the ⇖ at a location on the left border of the Formula Node.

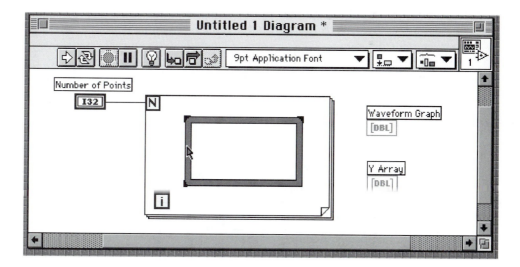

Then pop up on this border and select **Add Input** from the menu that appears.

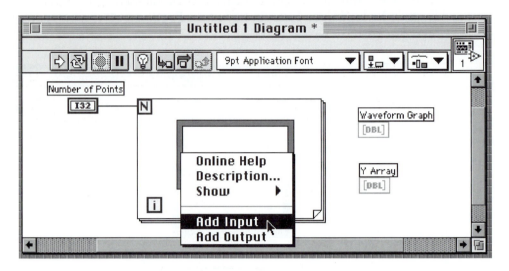

An indicator for this input will appear in the form of a small rectangular box perched on the Formula Node's border. The interior of this box will be highlighted, awaiting your character-based name for this equation variable. If this highlighting disappears (due to an inadvertent click of the mouse), use the ⟨🖑⟩ to restore access to the input box's interior. You can be as creative as you like in naming your input variable. LabVIEW allows Formula Node variable names to be of unlimited length, and the names are case sensitive. Long names, of course, have the disadvantage of occupying considerable diagram space. To conserve valuable diagram real estate, give the newly created input variable an uncreative short name—something like i. Secure this choice by pressing the Numeric Keypad's *<Enter>* key or clicking the mouse on an empty region of the block diagram.

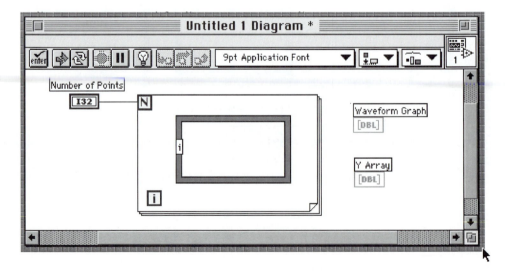

Wire the For Loop's iteration terminal to the Formula Node's input variable *i*. Then pop up on the Node's right border and select **Add Output** from the menu. Name this output variable *Y*. Note that an output variable is distinguished from an input variable by its thicker box.

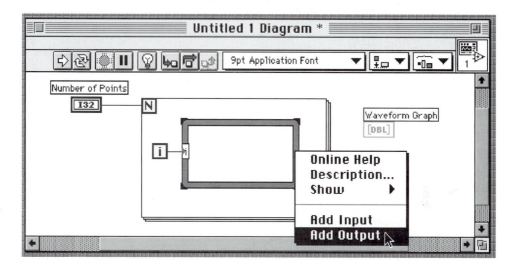

Now, using the 🖑, enter the equation *Y = sin(i/5.0);* into the interior of the Formula Node. Remember, all equations within a Formula Node must be terminated by a semi-colon (;).

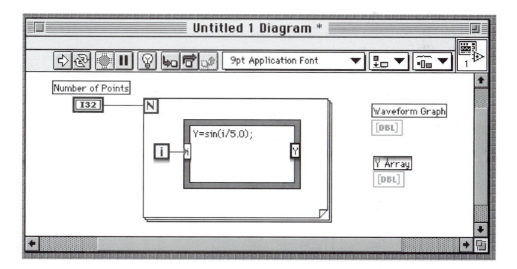

Finally, wire the Formula Node's *Y* output, through a tunnel at the For Loop border, to the Waveform Graph's terminal as well as the **Y Array** indicator's terminal. Then use the ↖ to reposition and resize all the objects, if desired.

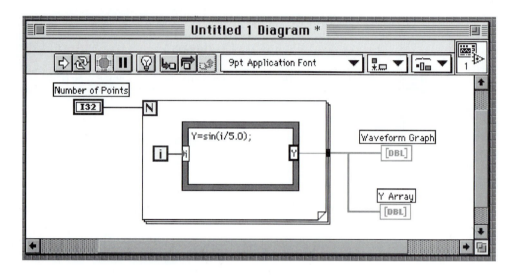

Debugging with Error List

Let's digress for a moment and explore one of LabVIEW's handy program debugging features. Intentionally put an error in your block diagram by deleting the semicolon within the Formula Node. Use the ✍ to highlight this character, then erase it with the *<Delete>* key.

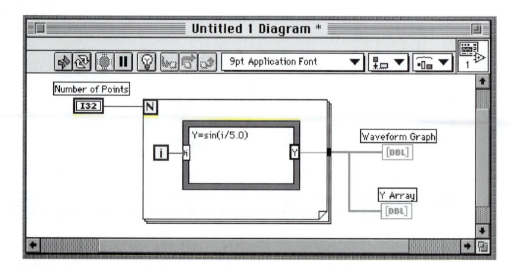

Now scrutinize the toolbar. There you will find that the **Run** button appears broken 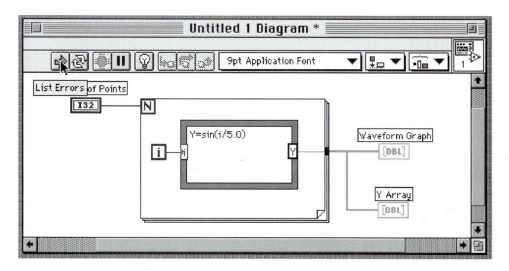, which is LabVIEW's way of communicating that an error exists within your program. But this indicator isn't the only help available.

Click on the (a tip strip alerts you that this action will produce a list of program errors). A dialog box will appear, listing all of the errors in your program. In the present case, just one error exists and is identified as **Formula Box:missing semicolon**.

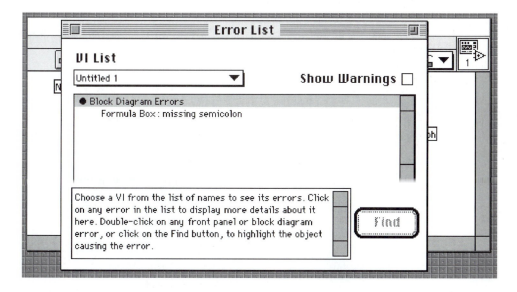

There's even more help to be found by clicking on the error comment **Formula Node:missing semicolon**. With this click, the lower box gives a detailed description of your problem. Clicking on the **Find** button will take you to the error's location on the block diagram, where you can then make the required repair. With all of this available online help, quite commonly a LabVIEW program can be debugged without having to flip through a user's manual.

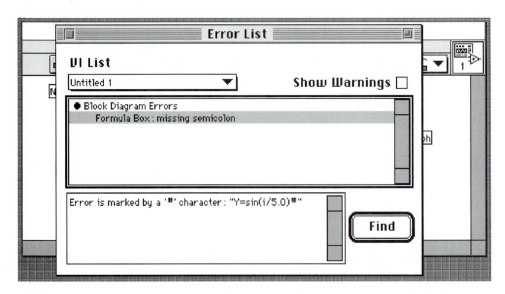

Return to the front panel after replacing the missing semicolon on your block diagram. Using the 🖑, set the **Number of Points** control to *100*, then run your program. It should produce a few beautiful cycles of the sine function, as shown. You may then view the actual values of the sine function using the Array Indicator.

Save this VI under the name **Sine Wave Generator-Formula Node** in the **YourName.llb** Library.

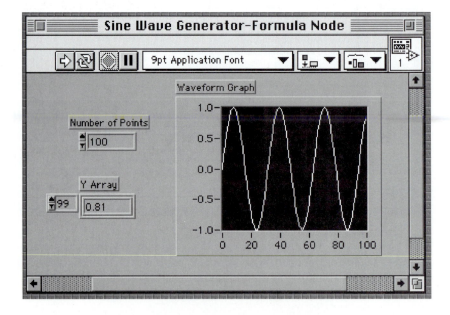

SINE WAVE GENERATOR USING A FORMULA NODE AND XY GRAPH

In this section, you will write a VI that generates a sequence of arguments X and their associated values of Y=sin(X) using a **For Loop** and **Formula Node** structure. The VI will then plot the resulting sequence of (X,Y) data points on an **XY Graph**. This program behaves as a sine-function generator and can be easily modified to generate any other arbitrary function.

The **XY Graph** is a general-purpose graphing option that produces Cartesian-style plots. That is, each point on the plot is located by its (X,Y) value. The entire set of (X,Y) values is presented to this routine in the form of a cluster consisting of two 1D arrays of numerical values. The first and second array in the cluster correspond to the sequence of X and Y values, respectively.

Place a labeled **XY Graph** (from **Controls>>Graph**) on a new front panel, along with two Array Indicators labeled **X Array** and **Y Array**. Also place a **Digital Control** labeled **Number of Points** on the panel. Change this control's data-type to **I32**.

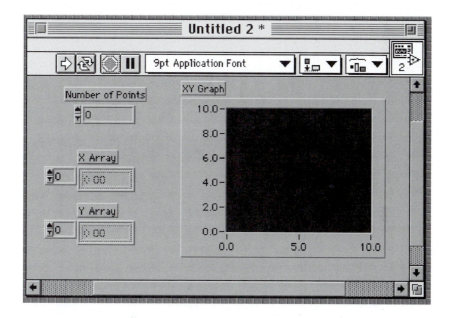

To produce the two required arrays containing the waveform's X- and Y-coordinates, write the following block diagram. Take care to terminate each equation with a semicolon, and remember that the characters are case sensitive (that is, the capitalized version of a particular letter is distinct from its lowercase rendering).

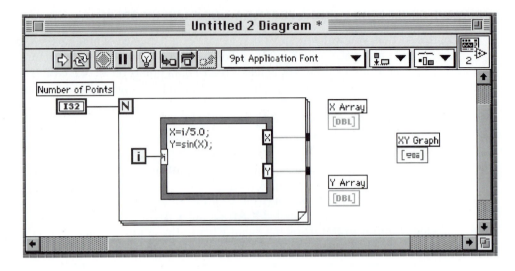

Creating an XY Cluster

To determine what type of input the **XY Graph** requires, activate its Help Window by placing the mouse cursor over the XY Graph's terminal [⌐₀₆] and pressing *<Cmd><H>* (Macintosh) or *<Ctrl><H>* (Windows). The Window will appear as shown.

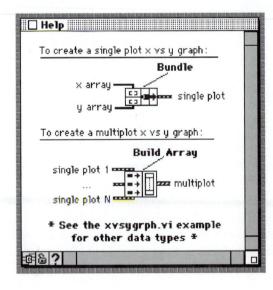

This Help Window indicates that to plot a single waveform, the **XY Graph** expects a cluster as input. This cluster is to contain the waveform's X and Y arrays and is formed using the **Bundle** icon. Let's call this bundle of a waveform's X and Y arrays its *XY cluster.*

For future reference, the Help Window also tells us that N waveforms can be plotted simultaneously on an **XY Graph**. In this situation, the data are passed to the Graph's terminal in the form of an "array of clusters." To form this object, one must first form the XY cluster for each of the N waveforms as above. Then, using the yet-to-be-studied **Build Array** icon, an N-element array is constructed with each element being a particular waveform's XY cluster.

Refer to your block diagram. You will find that the XY Graph's icon [⌐□□] presently denotes that, in the most general case of multiple waveform plotting, the data input is an array of clusters. The icon's square brackets denote the array nature of this data, the varying-size rectangles within signify that each array element is a cluster.

Complete the block diagram wiring as shown below. Use the **Bundle** icon (found in **Functions>>Cluster**) to form the sine wave's XY cluster. Note that when this cluster is wired to the XY Graph's terminal, the icon changes from [⌐□□] (array of clusters) to [□□□] (cluster). This transformation reflects the fact that your request for single waveform plotting requires the input of only a single cluster, rather than the array of clusters necessary for the more complex multi-waveform plots.

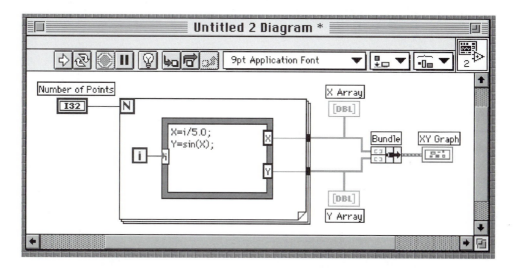

Return to the front panel, choose a value for **Number of Points**, then run your program. Marvel at the waveform you have created and reflect on how easy it would be to create other shapes by simply changing the equations within the block diagram's Formula Node. Slightly modified versions of this VI will prove very useful as data simulators in future studies.

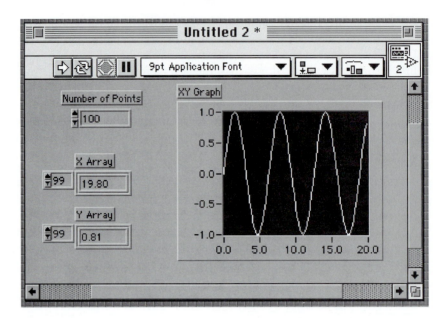

Sine-Wave Data Simulator

Let's modify the above program to create one of these data simulation VIs now. In this VI, we will assume a physical system under investigation is producing a pure sine-wave signal Y of frequency f and amplitude A. Further, we will assume that our experimental instruments measure ("sample") this signal Y at N equally spaced times X, where Δt is the time interval between one sample to the next. Mathematically, the resulting data set will be given by the following two relations:

$$X = i \, \Delta t$$
$$Y = A \sin(2\pi f X) \tag{1}$$

where i = 0, 1, 2, ..., N − 1.

Our plan is to write a software program that creates the data set described by Equation (1) and thus simulates the (X,Y) output of the above-mentioned experiment. To accomplish this task, add a **Digital Control** to your front panel and label it **dt**. Pop up on this control, select **Format & Precision**, then change the value of **Digits of Precision** from its default value of *2* to *6*. Use the ᛘ to resize the **Digital Control**, so that all of its digits are visible, and to arrange the front-panel objects in some pleasing pattern. Finally, save this VI under the name **Sine Wave** in **YourName.llb**.

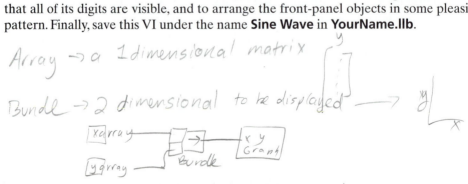

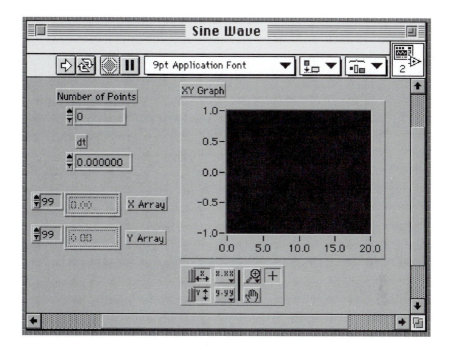

Switch to the block diagram and code the Formula Node to calculate Equation (1) with A = 4.0 and f = 100 Hz. Note that the Formula Node recognizes the characters *pi* as the mathematical constant π.

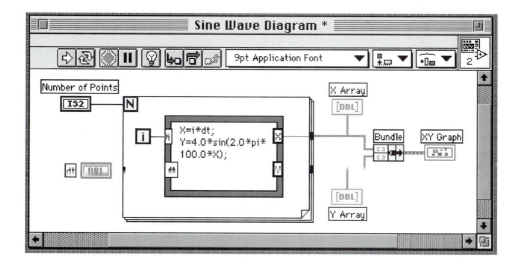

Now return to the front panel. Input *100* and *0.001* for **Number of Points** and **dt**, respectively, then run your program. With this choice of parameters, you are simulating an experiment that acquires data from t = 0.000 to t = 0.099 seconds at time increments of 0.001 seconds. Thus, for a 100 Hz sine-wave signal, one would expect to observe approximately 10 of its cycles on the XY Graph.

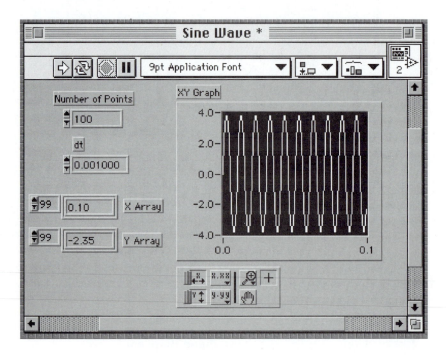

Try running your VI again with *100* and *0.0001* input for **Number of Points** and **dt**, respectively. Now the data-gathering times span the range from t = 0.0000 to t = 0.0099 seconds. Thus only one cycle of the 100 Hz signal will appear on the XY Graph, as shown.

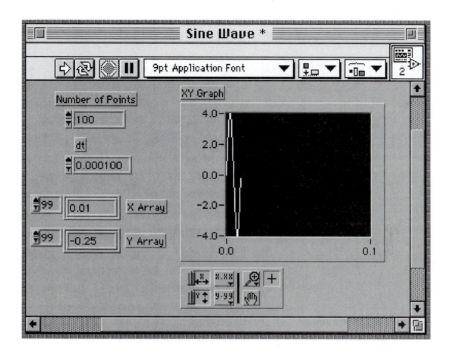

Formatting the Scale of a Plot Axis

Above, we have confronted a precision problem in the X-axis scaling. By default, LabVIEW plots provide only one digit of precision on axes scaling, meaning the smallest allowed scaling increment is 0.1. However, this setting can be modified either through the XY Graph's pop-up menu or the Palette. Here's how to effect this change using the Palette. Place the 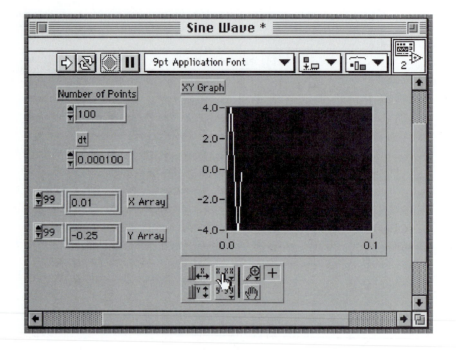 over the Palette's **X Scale Format** button, as shown below.

Click and select **3** from the **Precision** palette. You may wish to peruse the other options available in this menu for a moment.

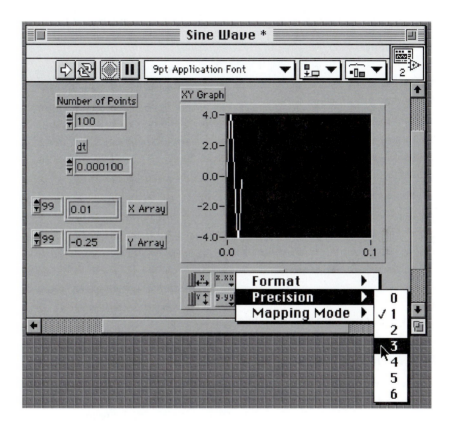

Now run the VI again and enjoy the results.

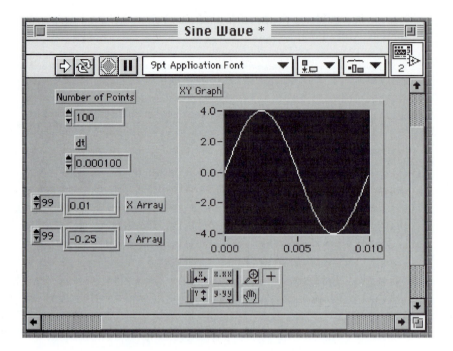

Save your work up to this point on **Sine Wave**.

CREATING AN ICON USING THE ICON EDITOR

A well-written LabVIEW program is hierarchical in nature. It consists of a top-level VI whose front panel accepts inputs and displays outputs, while its block diagram is constructed from lower-level subVIs. These subVIs, which are analogous to subroutines in a text-based language such as C, may call even lower-level subVIs, which in turn may call still lower-level subVIs. Just as there is no limit to the level of subroutine layering in a C program, there is no limit to the layers of subVIs used in a LabVIEW program. This modular approach to programming makes programs easy to read and debug.

Your program's subVIs either can be taken from LabVIEW's extensive libraries of built-in icons (found in the Functions Palette) or can be custom written by you. It is this latter point that we now wish to focus our attention upon. The importance of what you are about to learn is this: Any VI that you write can then be used as a subVI in the block diagram of a higher-level VI.

To use a program as a subVI, it must have an icon to represent itself in the block diagram of the higher-level ("calling") VI. There are two steps in creating this icon: designing its appearance and assigning its connectors. Let's learn these skills by creating an icon for your **Sine Wave** program.

Icon Design

First, here is the procedure for designing the icon's appearance. Position the mouse cursor (it doesn't matter which tool it is manifesting at the moment) over the *icon pane* in the upper right-hand corner of the front panel.

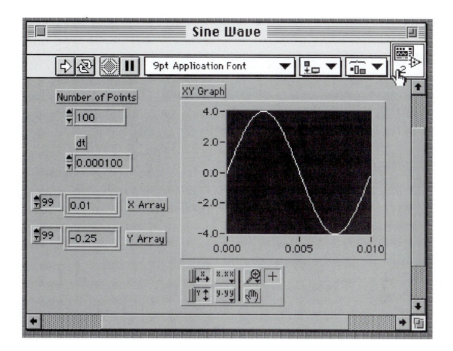

Pop up on the icon pane and select **Edit Icon** from the menu.

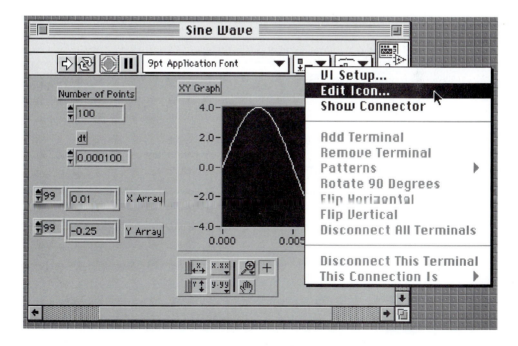

The *Icon Editor* window, shown in the following illustration, will appear.

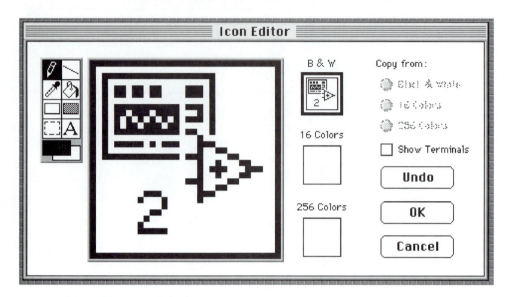

Within this window, you will find a default picture of the icon and a palette of tools that may be used to redesign its appearance. The tools in the palette, some of which may be familiar to you from computer-based drawing programs, have the following functions:

Pencil	Draws and erases pixel by pixel.
Line	Draws straight lines. Press <*Shift*> to restrict drawing to horizontal, vertical, and diagonal lines.
Dropper	Copies foreground color from an element in the icon.
Fill Bucket	Fills an outlined area with the foreground color.
Rectangle	Draws a rectangular border in the foreground color. Double click on this tool to frame the icon in the foreground color.
Filled Rectangle	Draws a rectangle, bordered with the foreground color and filled with the background color. Double click to frame the icon in the foreground color and fill it with the background color.
Select	Selects an area of the icon for moving, cloning, deleting, or other changes.
Text	Enters text into the icon design.
Color	Displays the current foreground and background colors. Click on each to get a palette from which to choose new colors.

You may wish to explore the use of these tools and create a sophisticated design for your icon. The following step-by-step description will result in an icon with only rudimentary features.

Place the mouse-cursor over the **Filled Rectangle** tool, then double click.

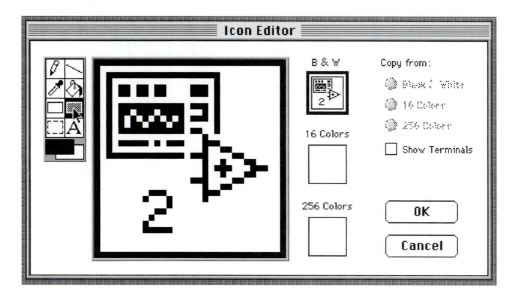

This action will frame the icon in the foreground color (default value is black) and fill it with the background color (white by default). You now have a framed blank canvas on which to create your icon design.

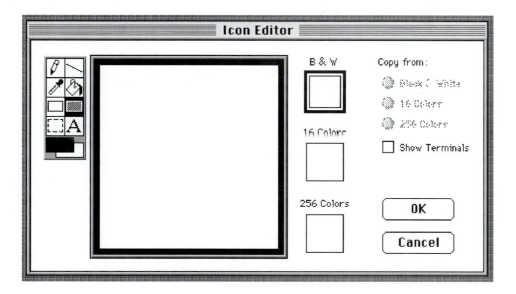

Next use the **Text** tool to enter **Sine Wave** in the icon's interior. Then click on the **OK** button to save the icon design.

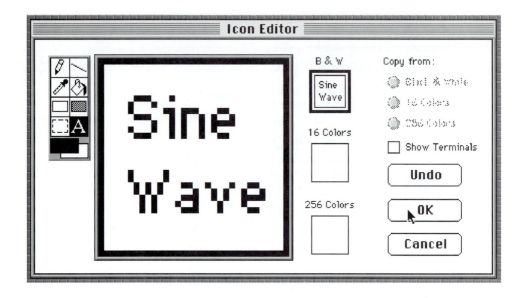

The Icon Editor window will close, returning you to **Sine Wave**'s front panel, with your new icon design now in the icon pane.

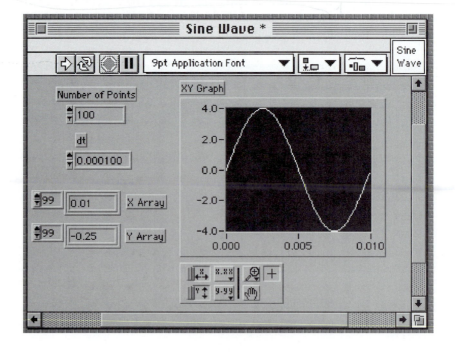

Connector Assignment

Second, here's how to assign the icon's connectors. Pop up on the icon pane and select **Show Connector**.

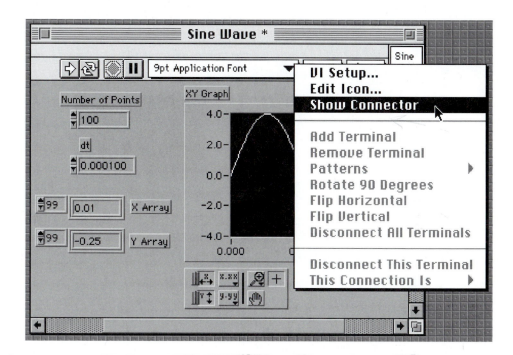

A connector consisting of a pattern of rectangular boxes will appear in the icon pane. Your job now is to associate each of your program's inputs and outputs to a particular box in this pattern. With the convention that inputs are on the left and outputs are on the right, LabVIEW has chosen a pattern based on the number of controls (**Number**

of Points and **dt**) and indicators (**X Array**, **Y Array**, **Graph's XY cluster**) on your VI's front panel.

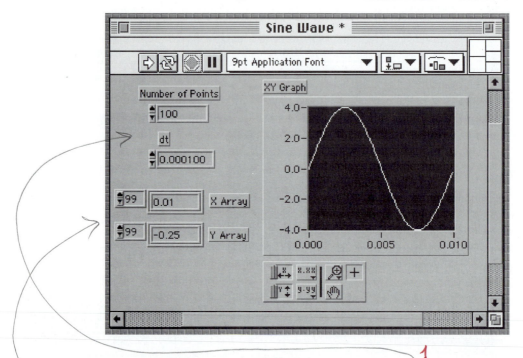

When we use **Sine Wave** as a subVI in the future, we will pass two quantities to it—the number N of data points we wish it to generate at a given sampling interval dt—and take two quantities from it—the X (time) and Y (sine-wave) arrays generated. Thus we require a connector with only two inputs and two outputs. To change the connector pattern, pop up on the icon pane and choose the desired connector from the **Patterns** palette.

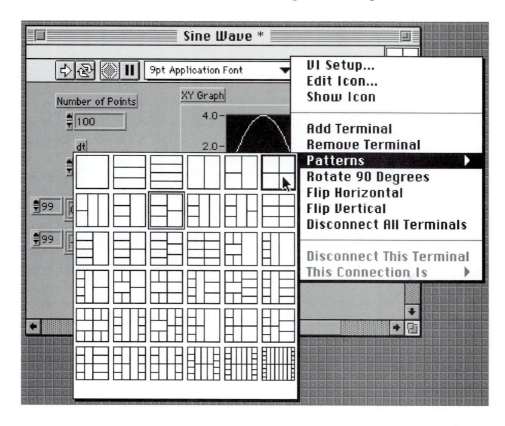

For future reference, if the exact input-output pattern you desire isn't directly on the palette, you can create it by first selecting a related pattern and then applying some operation from the menu such as **Flip Horizontal**. Or, you can create your own pattern of inputs and outputs by selecting **Add Terminal** and/or **Remove Terminal** in the pop-up menu (you might take a few moments to explore the operation of these two commands).

Now assign the **Number of Points** control to the pattern's upper left-hand terminal through the following steps. Click on the upper left-hand terminal.

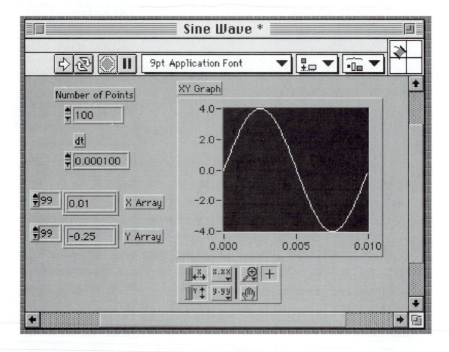

The cursor automatically changes into the Wiring Tool and the terminal turns dark as shown.

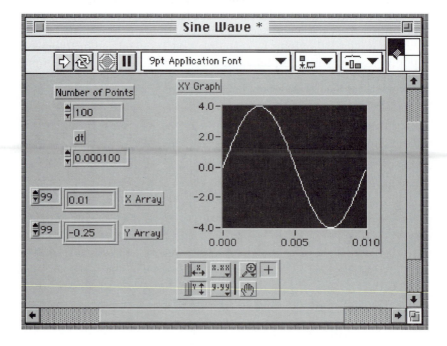

Click on the **Number of Points** control. A moving dashed line will then frame the control.

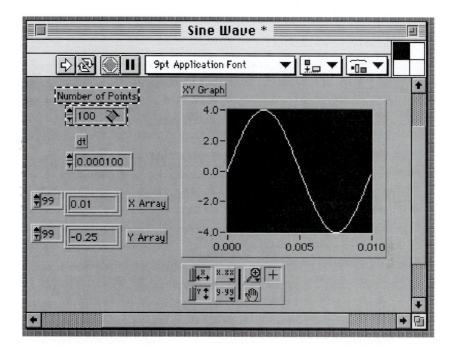

Next, click on an open region on the front panel. The dashed line will disappear and the selected terminal will dim, indicating that you have completed the assignment procedure. In a similar manner, assign the lower left-hand terminal to **dt** and the upper and lower right-hand terminals to the **X Array** and **Y Array** indicators, respectively.

(You only need to click on the open region of the front panel once, at the very end of this assignment sequence.) The icon pane will then appear as shown below.

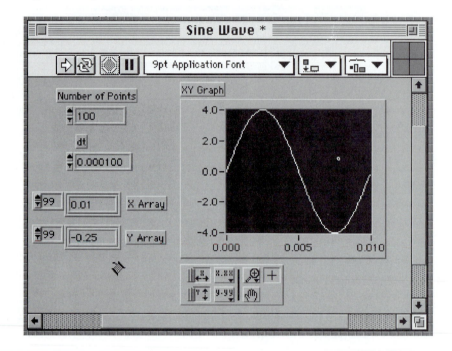

You may return the icon design to the icon pane by selecting **Show Icon** from the pop-up menu.

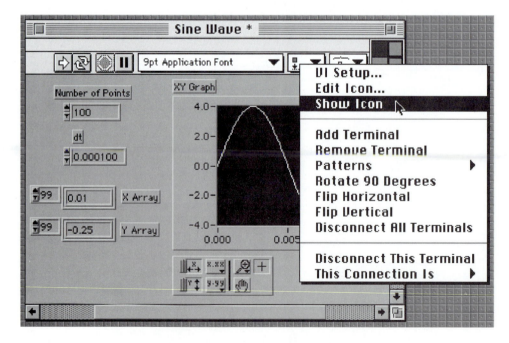

To demonstrate the success of your icon creation, place the mouse cursor over the icon pane and activate the Help Window.

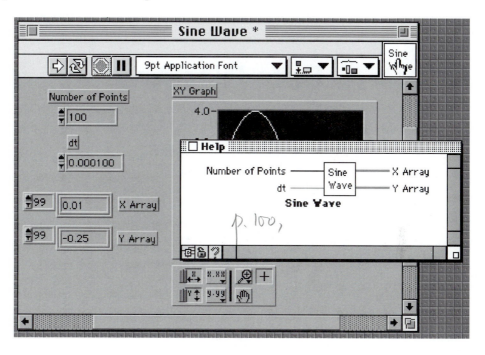

Congratulations, you've now completed your first custom-made VI. Using **Save** in the **File** menu, store this final version of **Sine Wave**. In the next chapter, you'll discover how to use a custom-made VI as a subVI in a higher-level program.

CHAPTER 4

Data Files And Character Strings

The outcome of any experiment is embodied in the data it generates. In computer-controlled experimentation, then, it is important to have the ability to save experimental data in disk files for future analysis. Hopefully these data are stored in a convenient format so that they can be easily read by a user-written program or a commercially available software package that analyzes and displays the experimental results.

Computers commonly communicate alphanumeric data by using the *American Standard Code for Information Interchange* (ASCII). In this coding scheme, seven bits of a byte are used to represent a character, while the byte's eighth "parity" bit is used for error checking when the data is received by a reader. The ASCII set of $2^7 = 128$ distinct states is used to represent all of the keyboard's alphanumeric characters as well as non-printable control characters such as carriage return (CR) and line feed (LF). In the Extended ASCII set, all eight bits of the byte are utilized for character coding, yielding 256 distinct states.

A sequence of ASCII characters is called a *string*. Character strings, of course, can be used to represent text messages. However, strings are also useful in passing commands and data to and from the computer and stand-alone instruments. Additionally, as you will find in this chapter, strings can be used to store numerical data in files on a computer disk.

The closest approximation to a universally readable data file format is the *ASCII text file*. Storing data in this manner has the following advantages. Your files will be read accurately by computers of all manufacturers. Additionally, the files can be viewed with a word processor and, if desired, easily cut and pasted into a document for report generation. Finally, your data will be easy to import into commercially available data analysis software packages. A large number of these application programs prefer that your file is in tab-delimited ASCII text, or what is commonly called the *spreadsheet format*. In this format, tabs separate columns and End of Line (EOL) characters separate rows as shown in the following figure.

0.00➔ 0.4258¶	
1.00➔ 0.3073¶	➔ = Tab
2.00➔ 0.9453¶	
3.00➔ 0.9640¶	¶ = EOL
4.00➔ 0.9517¶	

Opening this file using a spreadsheet program such as Microsoft Excel yields:

	A	B	C
1	0.00	0.4258	
2	1.00	0.3073	
3	2.00	0.9453	
4	3.00	0.9640	
5	4.00	0.9517	
6			

Alternately, numerical data may be stored as a *binary file*. This file format is simply a bit-for-bit image of the data that resides in your computer's memory. Thus, when data are exchanged between a binary file and computer memory, little or no data conversion is required so you get maximum performance. Also, binary data files provide the most memory-efficient storage method for numerical data. To demonstrate this fact, consider the number of bytes necessary to store the integer *54321*. Since this number is less than $2^{16} = 65536$, it will require only two bytes of memory in a binary-format file. However, in an ASCII text file, this number will occupy 5 bytes of memory, one byte for each of the characters *5*, *4*, *3*, *2*, and *1*. The disadvantage of binary files is their lack of portability. They cannot be viewed by a word processor and they cannot be read by any program without a detailed knowledge of the file's format.

LabVIEW contains built-in VIs that facilitate data storage and retrieval using either the ASCII text or binary file format. If ease of use and compatibility with a spreadsheet application program are of most concern, use the ASCII-based file storage. If, on the other hand, efficient memory use and high speed are desired, a binary file is the best choice.

In this chapter, you will learn how to store data in an ASCII-based spreadsheet file. You will write a program that generates data using your custom-made **Sine Wave** as a subVI and then store this data by implementing icons from **Functions>>File I/O**.

STORING DATA IN A SPREADSHEET-FORMATTED FILE

Open a new front panel. Name it **Spreadsheet Storage** and store it in **YourName.llb** using the **Save** command. Place a **Digital Control** (with Representation of **I32**) on the panel and label it **Number of Points**.

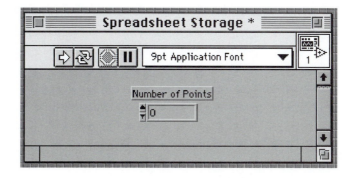

Now switch to the block diagram, where you'll find the terminal for the **Number of Points** control. Our plan is to use this control to tell your **Sine Wave** program to generate a certain quantity of data points, which you will subsequently store in a spreadsheet data file.

Placing a Custom-Made VI on a Block Diagram

First, you must place **Sine Wave** (as a subVI) on your block diagram. On the Functions Palette, choose the **Select a VI**...subpalette.

A dialog window will appear. Locate and open the **YourName.llb** Library.

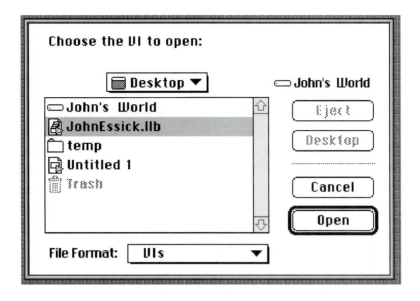

When the **File Dialog** window appears, double click on **Sine Wave**, or highlight it and then click on the **OK** button in the dialog box.

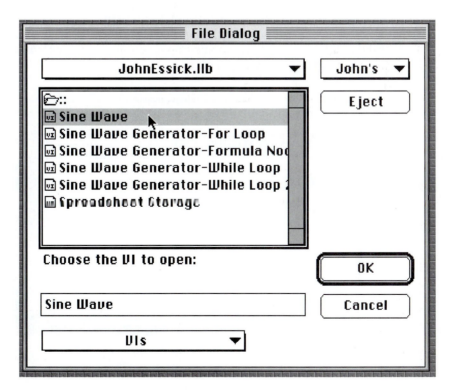

You will then be returned to the block diagram, where you can place the custom-written icon in a convenient position. If you haven't changed anything in the **Sine Wave** VI since the previous chapter, it should be programmed to calculate a 100 Hz sine wave of amplitude 4.0.

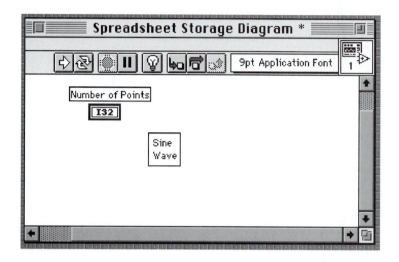

Wire the front-panel control's terminal to the **Number of Points** input of the **Sine Wave** icon. Then select a **Numeric Constant** from **Functions>>Numeric**, insert the value *0.001* and wire it to Sine Wave's **dt** input (alternately, you can pop up on the **dt** terminal and select **Create Constant**). **Sine Wave** will then calculate an array of sinusoidal values at 0.001 second increments. Pop up on the **Numeric Constant** and select **Show** to make its label visible, if desired.

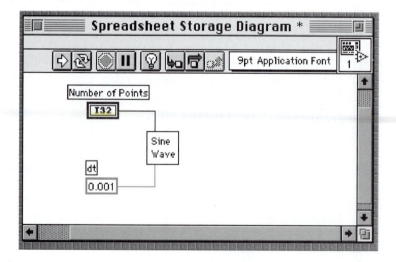

Using **Functions>>File I/O**, place a **Write To Spreadsheet File.vi** icon on the diagram.

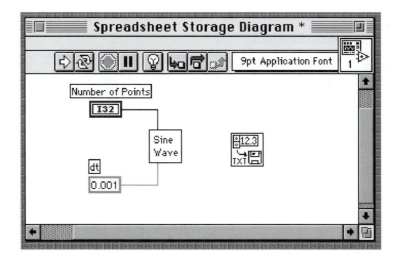

Storing a 1D Data Array

To understand how the **Write To Spreadsheet File.vi** icon functions, view its Help Window, as shown below. Here, you will find that this VI has the potential to function in numerous ways. Also, as is generically true for Help Windows, the default values for the icon's various inputs are given in parentheses. In its most basic operating mode, one simply provides a 1D array of numerical values to Write To Spreadsheet File.vi's **1D data** input. The VI then prompts the user for a file name and stores the values under this title using the spreadsheet format. By operating **Write To Spreadsheet File.vi** in this elemental way, you will gain insight into the necessity of its other available options.

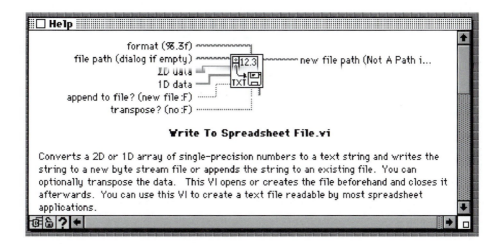

Wire Sine Wave's **Y Array** (sine-wave values) output to the Write To Spreadsheet File.vi's **1D data** input.

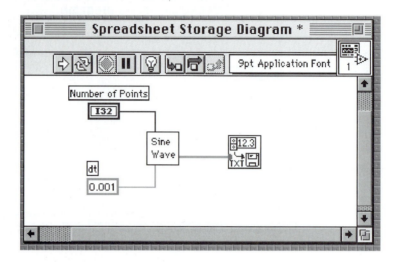

Now return to the front panel, input a small number such as *10* for **Number of Points**, then run your program.

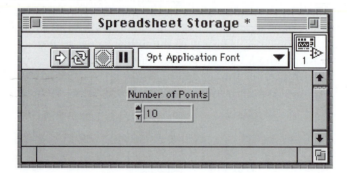

A dialog box called **Choose file to write** will appear. Navigate to the directory (or folder) in which you would like the spreadsheet file to be stored. In the next illustration, the folder called **Data** on the hard drive named **John's World** is being selected on my Macintosh system.

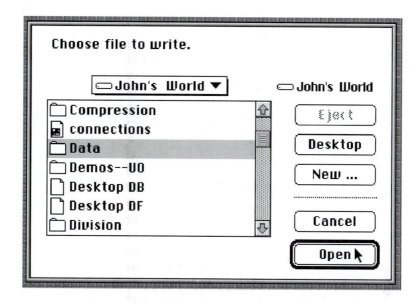

If you are using a Macintosh system, you must next press the **New...** button.

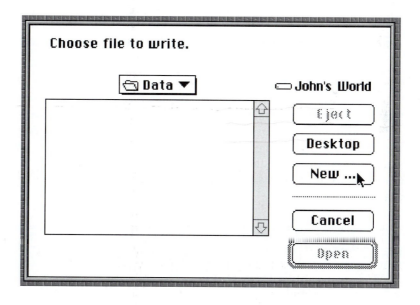

Then, on all platforms, in the dialog box requesting a name for the spreadsheet file about to be produced, name the file something extravagant like **Sine Wave Data**, unless you're a Windows 3.1 user. The eight-character length limit for file names in Windows 3.1 will only permit something like **SineWave.dat**. Once properly named, press the **File** (Macintosh) or **Save** (Windows) button.

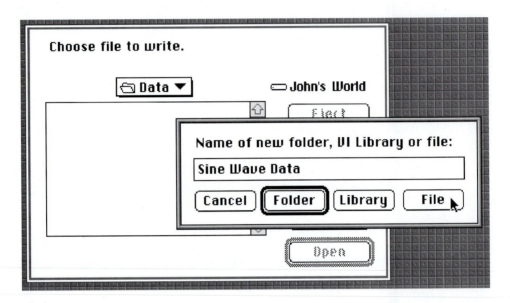

The program will then complete execution by creating the requested file.

Use a word processor program to open the ASCII-based **Sine Wave Data** (or **SineWave.dat**) spreadsheet file. Shown next, is what I found by viewing this data file in Microsoft Word (the appearance of your file may vary slightly due to the particular tab settings of your word processor). Here I have activated Word's **Show ¶** command, which allows one to see the file's non-printing characters such as Tab (⬧) and EOL (¶). If your word processor has a similar option, use it to view these usually hidden characters.

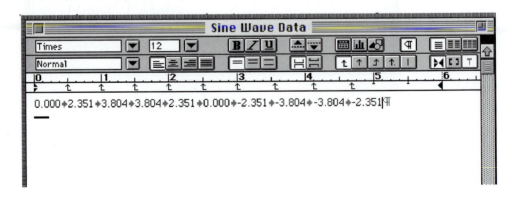

We see that the 10 sine-wave values are delimited by tabs and that the sequence concludes with an EOL. From our knowledge of the spreadsheet formatting convention, we conclude that a spreadsheet application program will interpret this string as a row of data, placing these 10 numerical values in a sequence of columns contained within a single row. Close **Sine Wave Data** with the word processor, but don't save it (or else the program will possibly embed its own characteristic formatting statements within the file).

If available, use a spreadsheet application program to read your **Sine Wave Data** file. To do this, your application program may give you the option of using an **Open** or an **Import** command. Either will work. You may have to tell the program that you're reading in a text file (as opposed to binary, Excel, etc.). Also you may encounter a dialog window in which you must tell the program that your data is tab-delimited. Next, I show the result of reading the **Sine Wave Data** file using Microsoft Excel. As expected, the 10 numerical values appear in 10 sequential columns of a single row.

	A	B	C	D	E	F	G	H	I	J	K
Sine Wave Data											
1	0.000	2.351	3.804	3.804	2.351	0.000	-2.351	-3.804	-3.804	-2.351	
2											
3											
4											
5											
6											
7											
8											
9											
10											
11											
12											
13											

Transpose Option

If you have a bit of experience with spreadsheet applications, the above illustration will cause some concern. In using such programs to generate plots, perform curve-fitting and other useful data analysis operations, the array of values for a particular quantity (e.g., sine-wave displacement) is expected to reside in a single column with the array's elements indexed by the row numbers. This organizational scheme for data is often called *column-major order*. We have seen that, in its default setting, the **Write To Spreadsheet File.vi** icon does not function in a manner consistent with this convention. However, once this problem has been identified, it is trivial to remedy by consulting this icon's Help Window. Here you will find that the VI possesses a **transpose?** input, whose default value is FALSE. By wiring a TRUE **Boolean Constant** to this input, the file will be recorded in a "column-like" manner, rather than "row-like." Perform this wiring on your block diagram. An easy way to obtain the **Boolean Constant** is to pop up on the **transpose?** input. Remember that the value of the **Boolean Constant** can be switched by using the Operating Tool.

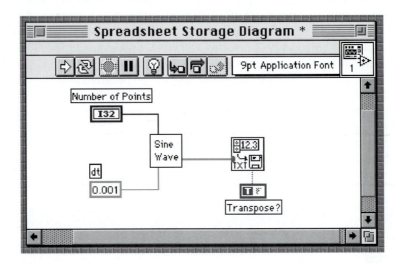

Run your program to produce a spreadsheet file of 10 sine-wave values. You can either reuse the file name **Sine Wave Data** (by opening this name in the **Chose file to write** dialog box) or else invent a new name. View this file using a word processing program. As shown below (using Microsoft Word), we see that each data value is separated from its neighbor by an EOL character. Thus, when read by a spreadsheet application, this array of values will be placed in a single column.

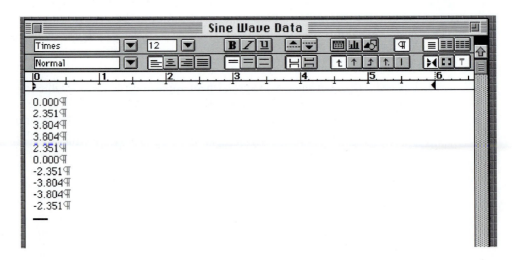

If available, open the data file in a spreadsheet application. Here (using Microsoft Excel), we see that our expectation of column-like data is fulfilled.

	A	B	C	D	E	F	G	H	I	J
Sine Wave Data										
1	0.000									
2	2.351									
3	3.804									
4	3.804									
5	2.351									
6	0.000									
7	-2.351									
8	-3.804									
9	-3.804									
10	-2.351									
11										
12										
13										

Storing a 2D Data Array

To analyze and/or plot **Sine Wave**'s sine-wave data using a spreadsheet application, one would need to import its X Array and Y Array [corresponding to Time X and $\sin(2\pi fX)$, respectively, where f is the frequency] into two spreadsheet columns. Let's see how, by implementing the **2D data** input of **Write To Spreadsheet File.vi**, it is easy to modify our **Spreadsheet Storage** program to produce the appropriate data file. We start with **X Array** and **Y Array**, the two 1D array outputs from **Sine Wave**. Each of these 1D arrays has N elements, where N is determined by the value of the **Number of Points** control. We now wish to splice these two 1D arrays together to form a 2D array with the N values of **X Array** and **Y Array** in the 2D array's first and second row, respectively. Mathematicians would call this object a 2-by-N matrix; in LabVIEW it is called a 2D array.

To construct this 2D array, you will use an icon called **Build Array**. However, first you must open up some room for **Build Array** between **Sine Wave** and **Write To Spreadsheet File.vi**. To widen the gap between these two icons, click on the wire that connects them, then delete it. Next, place the Positioning Tool as shown here.

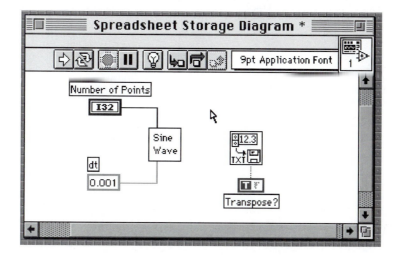

Click, then drag the cursor until a dotted rectangle frames both the **Write To Spreadsheet File.vi** and the **Boolean Constant** icon.

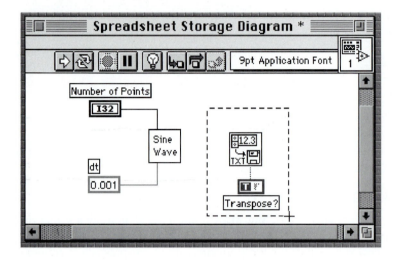

When you release the mouse button, both icons will be highlighted with a marquee. You can then move them to the right as one object by either using the keyboard's *<Right Arrow>* key, dragging with the Positioning Tool while depressing the *<Shift>* key (allowing only horizontal motion), or simply dragging with the Positioning Tool.

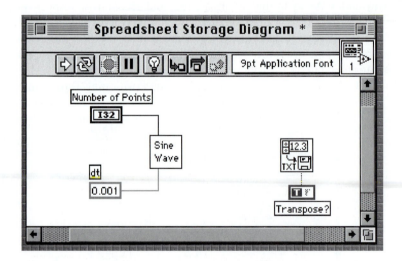

Now select the **Build Array** icon from **Functions>>Array**.

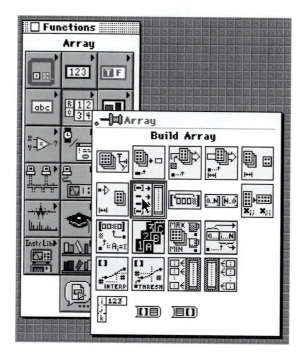

When you put this icon on the block diagram, it will initially appear with only one input.

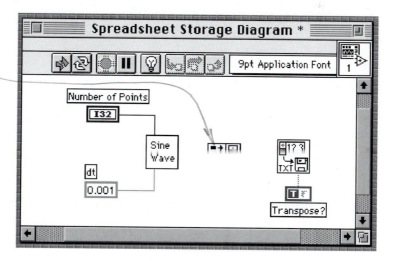

Place the Positioning Tool at one corner of the icon until it morphs into a ⌐. Resize the icon so that it can accommodate two inputs.

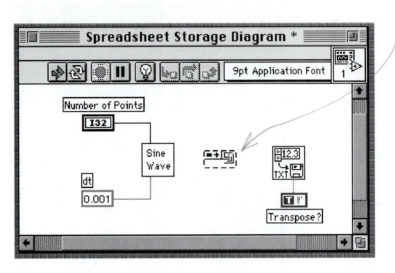

Then wire Sine Wave's **X Array** and **Y Array** outputs to the top and bottom inputs of **Build Array**, respectively. **Build Array** will splice these inputs together to form the required two-dimensional array.

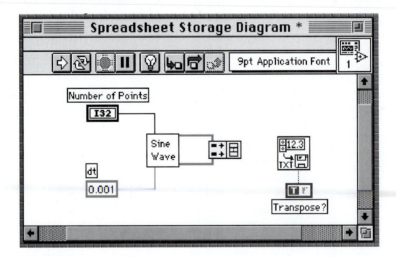

Wire the output of **Build Array** to the **2D data** input of Write To Spreadsheet File.vi. Note that this connection appears as a double wire, the LabVIEW designation for a 2D array.

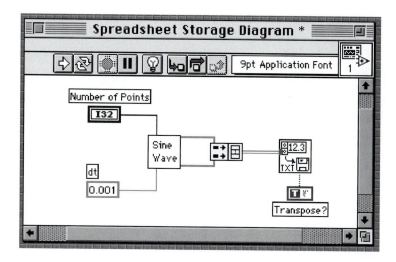

Return to the front panel and run your program with a small value such as *10* for **Number of Points**. Then view your resulting data file using a word processor. It should appear like this.

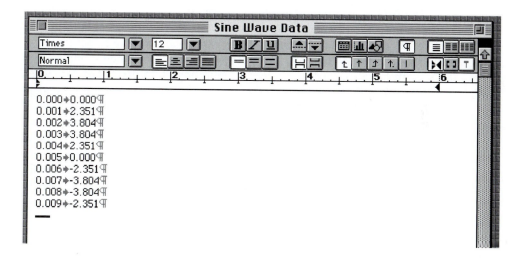

This formatting appears correct for importing the X Array and Y Array data into two adjacent columns of a spreadsheet. To verify this expectation, open your file with a spreadsheet application program. It should appear as below. If you're familiar with the operation of the spreadsheet application, you may enjoy plotting column B vs. column A or curve fitting the column B data to a sine function.

	A	B	C	D	E	F	G	H	I	J	
						Sine Wave Data					
1	0.000	0.000									
2	0.001	2.351									
3	0.002	3.804									
4	0.003	3.804									
5	0.004	2.351									
6	0.005	0.000									
7	0.006	-2.351									
8	0.007	-3.804									
9	0.008	-3.804									
10	0.009	-2.351									
11											
12											
13											

Controlling the Format of Stored Data

Now let's explore some of **Write To Spreadsheet File.vi**'s other options. First, you may have noticed that, when viewing your data file, each of its numerical values was stored with a precision of three digits to the right of the decimal point. Is this the precision of the numerical values produced by the **Sine Wave** VI?

Try this test: Open **Sine Wave** in **YourName.llb** using **Open**… in the **File** menu. Pop up on the **X Array** and **Y Array** front-panel indicators, select **Format & Precision**, then increase **Digits of Precision** from its default value of *2* to something larger like *8*. You may then want to resize the indicators so that all of the new digits are visible. Run the VI with some appropriate choices for **Number of Points** and **dt**. You will then find that the Array Indicators display values of the precision that you requested. Now, with this modification in **Sine Wave**, rerun **Spreadsheet Storage**, then view the new **Sine Wave Data** file it produces. You will find that the stored values there still have a precision of three digits. Thus, we surmise that **Sine Wave** is not influencing the formatting of values within **Sine Wave Data**. In fact, independent of its **Digits of Precision** setting, the **Sine Wave** VI calculates highly accurate floating-point values, which it outputs to connecting wires. The **Digits of Precision** parameter merely selects how many decimal places of these highly accurate values are to be displayed on the VI's front-panel indicator.

Since we have eliminated **Sine Wave** as a suspect, we must look elsewhere within **Spreadsheet Storage** to find where the decision is being made to truncate the **Sine Wave**-produced values at the thousandths place upon storage in the file. A quick glance at Write To Spreadsheet File.vi's Help Window, where default input values are shown in parentheses, identifies that the decision is made at this icon's **format** input. By leaving this input unwired, we have instructed the icon to store numerical values in the default format of *%.3f*. To understand this instruction, we need to know that the **format** string obeys the following syntax:

% [WidthString] [.PrecisionString] ConversionCharacter

where optional features are enclosed in square brackets. Each syntax element, and how it affects the form of the stored number within the data file, is explained in the following table.

%	Character that denotes the beginning of a format specification
WidthString (Optional)	Integer specifying the total number of ASCII characters to be used to represent the stored number. If your specified value of **WidthString** exceeds the number of characters actually required to represent a particular number, the excess portion of the string will be padded with spaces. If **WidthString** is not specified (or the stored number requires more characters than the specified **WidthString**), the numeric string will expand to as long as necessary to represent the stored number.
.PrecisionString (Optional)	Period (.) followed by an integer specifying the number of digits to the right of the decimal point in the stored number. If **.PrecisionString** is not specified, six digits to the right of the decimal point are stored. If only the period (.) appears and **PrecisionString** is missing or zero, no digits to the right of the decimal point are stored.
ConversionCharacter	Single character that specifies in what manner the number is to be stored, with the following code: d decimal integer x hex integer o octal integer f floating-point with fractional format e floating-point with scientific notation

Thus we see that *%.3f* instructs **Write To Spreadsheet File.vi** to store numbers in fractional floating-point format, with each number's total width auto-adjusted under the restriction of maintaining three digits to the right of the decimal point.

Place a **String Constant** on the block diagram, define it to be *%8.5f*, then wire it to Write To Spreadsheet File.vi's **format** input. The easiest way to create the **String**

Constant, of course, is to pop up on the **format** input. You can also find this icon in **Functions>>String**.

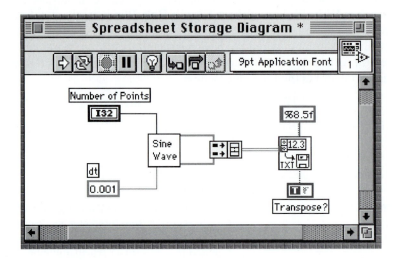

Run your program and store the data in a file. View the file using a word processor. Do the file's numerical values appear in the format that you specified? You might try exploring the effect of changing the **ConversionCharacter** to *d* (decimal integer) or *e* (floating point with scientific notation).

The Path Constant and Platform Portability

Next, let's look more closely at the act of naming the data file. We note that, in some cases, it would be desirable to hardwire the name of a data file into our program code. Glancing back at the Write To Spreadsheet File.vi's Help Window, we find that this option is available via the **file path (dialog if empty)** input. Previously, we have left this input unwired, thus the default **dialog if empty** (i.e., a dialog box prompting us for a file name) has been operational. Now we will take the alternate approach of programming-in a specific file name.

As you probably know, the path to a particular file in your computer system is specified by a hierarchical directory structure. The path is typically denoted by a **drive-name,** followed by **folder** (or **directory**) **names**, followed by the **filename** itself. On a Macintosh, these various levels of the hierarchy are separated by colons (:), while in Windows, backslashes (\) are used. Thus, on a Mac with a hard drive named **John's World,** the file **Sine Wave Data** within the **Data** folder has the pathname **John's World:Data:Sine Wave Data**. On a Windows machine, a similarly named file in the **Data** directory of the hard drive has the path **C:\Data\Sine Wave Data.dat**. The extension **.dat** is included to specify that the file contains data. To hardwire a specific file name into a program, one might anticipate that the pathname appropriate to your computing platform (Windows or Macintosh) is simply inserted into a **String Constant** and then this object is wired to the **file path** input of the **Write To Spreadsheet File.vi** icon. However, in deference to the issue of *platform portability*, this expectation is slightly modified.

Platform portability is one of the outstanding features of LabVIEW programming. This means, for example, that you can write a LabVIEW program called **Widget** on a Windows-based system, copy it to a floppy disk, then send it to a colleague for use in a Macintosh-based laboratory. Once your colleague has copied the contents of the Windows-formatted floppy onto his or her Mac system (with the help of a program such as PC Exchange), **Widget** can be opened from within LabVIEW running on the Macintosh system. While opening it, LabVIEW detects that **Widget** was written on another platform and recompiles it to use the correct instructions for the local processor. This whole process will, of course, be even easier if the file exchange takes place using e-mail. As seen above, one small part of this recompilation must include translating any hardwired path names that are contained within **Widget** into the form appropriate for the new system. To facilitate this translation process, a special LabVIEW object called the **Path Constant** (found in **Functions>>File I/O>>File Constants**) is used for the specification of path names. By dedicating the **Path Constant** to the sole use of enclosing path names, these special strings are distinguished from other character strings when the program is being ported.

Pop up on the **file path** input of **Write To Spreadsheet File.vi**, create a **Path Constant** on your block diagram, then enter a **pathname** appropriate to your computing platform into it. Here, I show the resulting diagram on my Macintosh system.

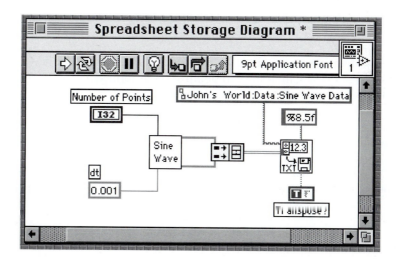

Run this program and verify that it performs as expected by viewing the resulting data file with a word processor or spreadsheet application program.

ADDING TEXT LABELS TO A DATA FILE

In the previous exercise, you created an ASCII text file with two tab-delimited columns of (X,Y) data, the first column containing a range of Times X in seconds, the next column containing the associated values of $Y = \sin(2\pi fX)$. Because of the file format, you were able to use a word processor or spreadsheet application program to view this file. A desirable addition to this data file would be the ability to provide descriptive text at the top of each column, labeling each experimental quantity. More generally, one might wish to include all sorts of text statements within the file that could provide a complete record of the experimental conditions and choice of parameters under which the data were taken. Let's see how to provide labels at the top of each data column. Once this skill is in place, it's a small step to include any other desired text into the file.

In our text-based method of data storage, the entire set of (X_i,Y_i) values is contained within one long ASCII character string. LabVIEW indexes these N values in the range of $i = 0$ to $N-1$, so the structure of this spreadsheet-formatted string is as follows:

$$X_0<Tab>Y_0<EOL>X_1<Tab>Y_1<EOL>\ldots X_{N-1}<Tab>Y_{N-1}<EOL>$$

One can then attach the labels *Time X* and *sin(2*pi*f*X)* as the initial entries in the first and second columns, respectively, by simply adding these text characters as a spreadsheet-formatted prefix *Time X <Tab> sin(2*pi*f*X)<EOL>*. The string will then appear as:

$$\text{Time X}<Tab>\sin(2*pi*f*X)<EOL>X_0<Tab>Y_0<EOL>X_1<Tab>Y_1$$
$$<EOL>\ldots X_{N-1}<Tab>Y_{N-1}<EOL>$$

Our task then is to construct the label prefix, append the long ASCII string containing all the data values, then store this entire string (termed a *byte stream*) in a file. Let's see how to do this.

In what follows, you will be adding some code in the upper region of the block diagram. The right scroll bar can assist in opening up some needed blank space in which to insert new program features.

First, construct the label prefix using the **Concatenate Strings** icon, which is found in **Functions>>String**. This icon splices together all its input strings into a single output string. Place **Concatenate Strings** on your block diagram. You will find that at first it has only two inputs.

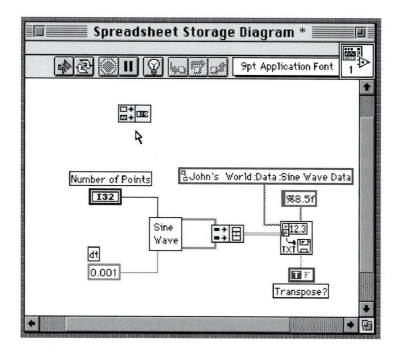

Place the Positioning Tool at one of the icon's corners, where it will morph into a resizing handle, then stretch the icon until it has four inputs.

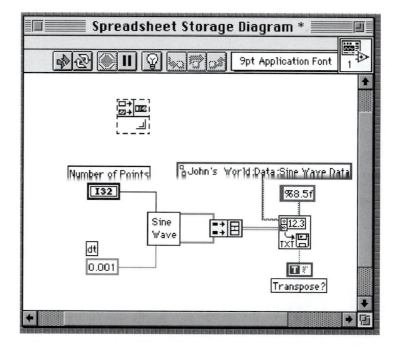

Wire a **String Constant** (found in **Functions>>String**, or by popping up on the input) containing the text *Time X* and *sin(2*pi*f*X)* to the first and third input of **Concatenate Strings**, respectively. Obtain a **Tab** and **End of Line** icon from **Functions>>String** and wire them to the second and fourth **Concatenate Strings** inputs, respectively.

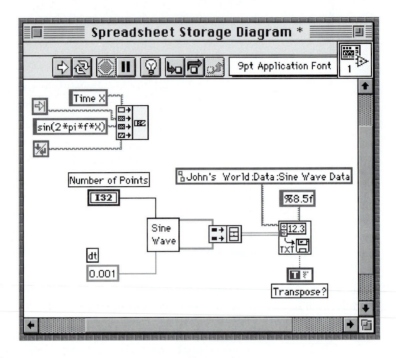

Next, place a **Write Characters To File.vi** icon on the diagram. This icon, found in **Functions>>File I/O**, can be used either to write a character string to a new byte stream file or to append the string to an existing file. The VI first opens (or creates) the file, writes the string, then closes the file. Its Help Window is displayed next.

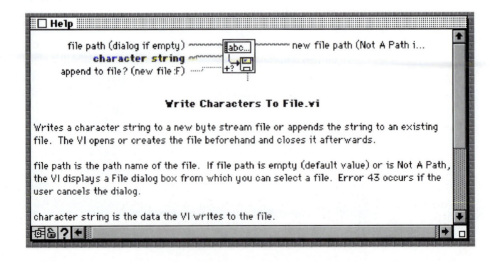

You will be using this icon to create a file and write its first entries—the column labels—so you may leave its **append to file?** input unwired. This input will then take on its default value of FALSE, signaling the icon to create a new file rather than to open an existing one. Since we are creating the file with this icon, disconnect the **Path Constant** from **Write To Spreadsheet File.vi** and rewire it to the **file path** input of **Write Characters To File.vi**. Alternately, you may leave the **file path** input unwired, if you prefer to provide a filename via the dialog box. Finally, wire the **Concatenate Strings** output to Write Characters To File.vi's **character string** input.

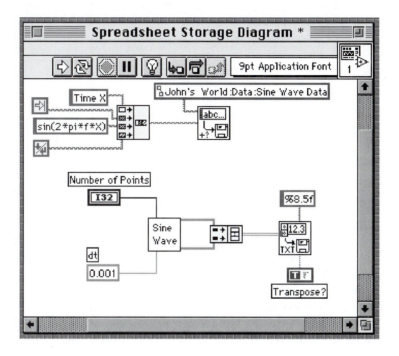

At this point, you may be wondering how the **Write To Spreadsheet File.vi** is going to know the name of the file to which you want it to write the spreadsheet-formatted data. In fact, you may be tempted to reconnect the **Path Constant** to Write To Spreadsheet.vi's **file path** input. However, because of the parallel manner in which LabVIEW icons execute, such a wiring arrangement could lead to the unfortunate situation of both icons attempting to write to the same file simultaneously. To avoid this potential conflict, we will take advantage of the principle of LabVIEW programming called *data dependency*. Simply stated, data dependency means that an icon cannot execute until data is available at *all* of its inputs. Take a look at the Write Characters To File.vi Help Window. There you will find that this icon has an output called **new file path**. Upon completion of writing operations to a file, the icon passes the file's **path-name** out of this terminal. Now here's a programming trick that will allow us to force **Write Characters To File.vi** to execute before **Write To Spreadsheet File.vi**. Simply wire Write Characters To File.vi's **new file path** output to Write To Spreadsheet File.vi's **file path** input. Then the **Write To Spreadsheet File.vi** icon will not operate until it receives data from the fully executed **Write Characters To File.vi** (as well as from the sub-diagram that produces the sine-wave arrays), assuring you that the icons will perform in the desired sequence.

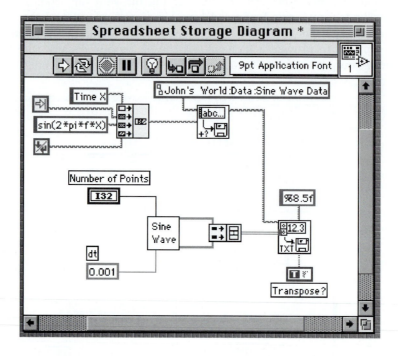

One final detail remains. Because **Write To Spreadsheet File.vi** is not creating a new file, but rather is opening an existing file and appending a character string to it, you must wire a TRUE Boolean Constant to its **append to file?** input. Get dual use from the **Boolean Constant** already on your diagram by the wiring arrangement shown next.

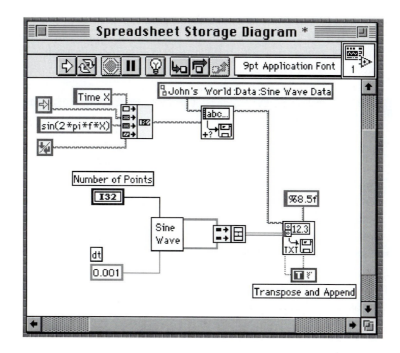

Run your program with whatever choice you'd like for **Number of Points**. Then view the resulting file using a word processor. With Microsoft Word, I found the desired labels atop each column just as we had planned. You may have to play with your word processor's tab settings to get the labels to align nicely with the numerical data.

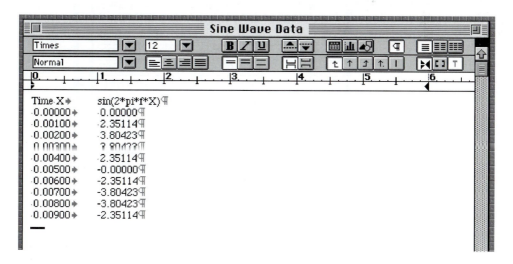

If available, try opening this text-based data file with a spreadsheet application program. Using the **Import** command in a spreadsheet program called KaleidaGraph, I first was presented with the following **Text File Input Format** window from which I selected the **Read Titles** option.

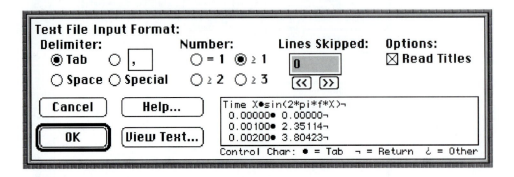

Upon pressing the **OK** button, the program produced the spreadsheet with labeled columns shown below, ready for analysis and graphing by the program.

	Time X		sin(2*pi*f*X)
0	0.00000		0.00000
1	0.00100		2.35114
2	0.00200		3.80423
3	0.00300		3.80423
4	0.00400		2.35114
5	0.00500		0.00000
6	0.00600		−2.35114
7	0.00700		−3.80423
8	0.00800		−3.80423
9	0.00900		−2.35114
10			
11			

Sine Wave Data

Backslash Codes

To conclude, let's explore a shortcut method for including nondisplayable ASCII characters within a **String Constant**. The trick is to pop up on the **String Constant** icon, then select **'\' Codes Display** (as opposed to **Normal Display**) from the menu that appears. When this option is activated, LabVIEW will interpret characters immediately following a backslash (\) as a code for nondisplayable characters in a manner listed in the following table. You must use uppercase letters for the hexadecimal characters and lowercase letters for the special characters such as Tab and CR.

'\' Code	LabVIEW Interpretation
\00–\FF	Hexadecimal value of an 8-bit character
\b	Backspace (ASCII BS, equivalent to \08)
\f	Formfeed (ASCII FF, equivalent to \0C)
\n	New Line (ASCII LF, equivalent to \0A)
\r	Carriage Return (ASCII CR, equivalent to \0D)
\t	Tab (ASCII HT, equivalent to \09)
\s	Space (equivalent to \20)
\\	Backslash (ASCII \, equivalent to \5C)

Let's demonstrate the convenience of backslash codes by employing them in the construction of the labeling-string *Time X <Tab> sin(2*pi*f*X) <EOL>*. To build this string, I must tell you that the EOL marker is platform-dependent. The definitions for EOL peculiar to each computing system are:

Macintosh: Carriage Return (\r)

Windows: Carriage Return, then New Line (\r\n)

Thus, by enabling the **'\' Codes Display** option, the entire column-labeling text can be constructed by inserting *Time X\tsin(2*pi*f*X)\r* within a single **String Constant**, assuming one is working on a Macintosh system. However, if in the future an attempt is made to port this program to another platform, the explicit expression of the EOL character will cause problems. Due to this portability issue, it is good programming practice to take the following slightly modified approach to the string construction: Enclose *Time X\tsin(2*pi*f*X)* within a single **String Constant**, then concatenate this object with the **End of Line** icon found in **Functions>>String**. The **End of Line** icon is a "smart" EOL character. At compile time, it checks which platform is currently being implemented and automatically generates the appropriate EOL character(s), thus infusing the program with portability.

Here is a step-by-step procedure for accomplishing the above. Start by replacing all of the labeling-string construction code from your **Spreadsheet Storage** diagram by a **Concatenate Strings** icon with two inputs. Wire a **String Constant** to the top input by popping up. Wire an **End of Line** icon from **Functions>>String** to the bottom input.

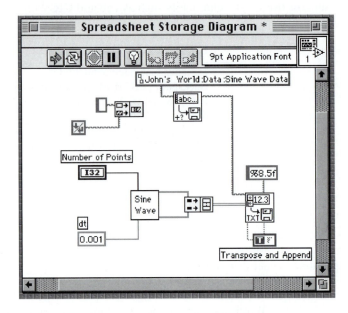

Pop up on the **String Constant** and enable backslash coding by selecting **'\' Codes Display** from the menu.

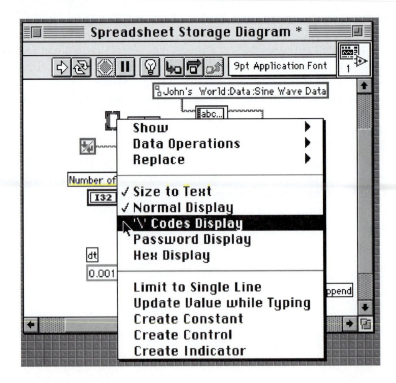

Using the Operating Tool, enter *Time X\tsin(2*pi*f*X)* into the **String Constant**. When you press the numeric keypad's *<Enter>*, LabVIEW will automatically replace any *<Space>* within your text with \s. When you are finished, you may have to resize and/or reposition this object using the Positioning Tool so that all of the enclosed text is visible. Then wire the **Concatenate String** output to the Write Character To File.vi's **character string** input.

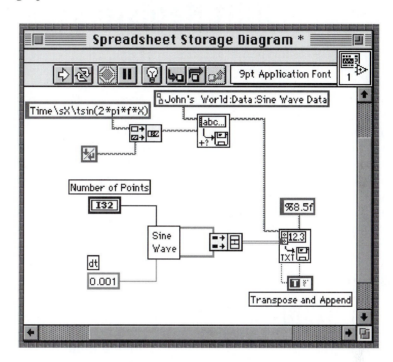

Run your program with some finite value selected for **Number of Points** and demonstrate that backslash coding produces a data file with the desired column labeling.

In closing this chapter, note that the above block diagram illustrates LabVIEW's color coding scheme for the complete range of its available data-types: floating point numbers (orange), integers (blue), Booleans (green), strings (pink) and file paths (aqua).

CHAPTER 5

Shift Registers

A looping structure allows one to implement a common requirement in computer programming—repeating the same operation many times. In earlier chapters, we found that LabVIEW provides two such looping structures, the For Loop and the While Loop. We discovered that, through the creation of a tunnel at its border, each of these loop structures possesses a form of memory. By toggling on the tunnel's auto-indexing feature, the loop (once it has completed its full execution) will output an array in which is stored the sequence of values created by the succession of loop iterations. Often, however, another form of memory is needed—that which interconnects successive loop iterations. In this guise, a value created by the previous iteration is transferred for use within the calculations of the present iteration. This form of memory is called a *local variable* and can be accomplished in LabVIEW's looping structures via the creation of a *shift register*.

Shift registers are created by activating a pop-up menu at a loop boundary. For example, you can create a shift register by popping up on a For Loop's border, then selecting **Add Shift Register**.

When you release the mouse button, the shift register will appear. It is comprised of a pair of terminals directly opposite each other on the vertical sides of the loop border.

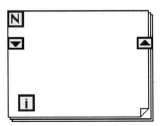

The right terminal stores a value upon completion of an iteration. This stored value is then shifted to the left terminal for use in calculations during the next iteration. Such a feature is useful, for example, in summing the components of a quantity calculated over the span of the entire set of loop iterations.

You can configure the shift register to remember values from more than one previous iteration. To accomplish this feat, pop up on the shift register's left terminal and select **Add Element** from the menu.

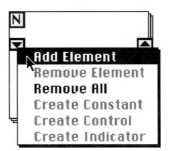

Upon this selection, a second left terminal will appear.

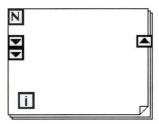

By repeating this operation, you can create as many left terminals as desired.

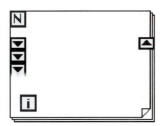

In the previous illustration, we have created three left terminals that will function as follows. When a subdiagram within the For Loop (not shown) is calculating some quantity during the ith loop iteration, the top-left terminal contains the value of this quantity calculated during the $(i-1)^{th}$ iteration, the middle-left terminal contains the value from $(i-2)^{th}$ iteration, the bottom-left terminal from $(i-3)^{th}$ iteration. Over the course of the complete For Loop execution, this set of left terminals will behave as a First In, First Out (FIFO) digital shift register, if that is a familiar concept to you.

In this chapter, you will explore several uses of local variables in LabVIEW programming. First, you will use shift registers to integrate a discrete data set numerically. Then, by recalling values from two past iterations, you will differentiate this same data set. Finally, you will see how to use an *uninitialized* shift register to create a subVI that may be used to share information between two or more calling VIs. This later form of memory is called a *global variable*.

ARBITRARY-FUNCTION DATA SIMULATOR VI

Imagine some physical system for which one of its measurable quantities Y (for example, temperature or pressure) is a function of the continuous variable X (e.g., position or time). That is, Y is perfectly described by $Y = f(X)$, where f is an analytic function. If an experimenter sets out to determine the function f, he or she will have to confront this reality: It is impossible to design an experiment that samples the quantity X and its associated Y values in a continuous manner. One must instead settle for discretely sampling X a total of N times and recording each associated Y value. Thus the experiment will result in a set of (X,Y) data points, where $X = X_0, X_1, X_2, ..., X_{N-1}$ and $Y = f(X_0), f(X_1), f(X_2), ..., f(X_{N-1})$.

In the next exercise, we will study the use of shift registers by writing a program that numerically integrates a discretely sampled data set. In an actual computer-controlled experiment, the discretely sampled data set might be obtained through the use of LabVIEW's built-in data acquisition (DAQ) VIs. Given this set, an integration VI might then be used to perform some appropriate analysis on the data. Because we haven't yet studied how LabVIEW can be used to record real data, we will use software to generate a simulated set of discretely sampled data. For our present purposes, it is actually beneficial to use simulated data. In contrast to a real experiment, we will know the exact functional form by which the data was generated. Thus we will be able to judge the accuracy of our numerical integration methods by comparison with an analytical result.

We desire a VI that, given a lower limit X_0 and upper limit X_{N-1}, creates N equally spaced points along an X-axis, then produces an associated array of $Y = f(X)$ values, where f is an analytic function. Your previously produced **Sine Wave** program provides a good starting point for writing this new VI. Open **Sine Wave** in **YourName.llb**, then clone it by selecting **Save As…** from the **File** menu. In the dialog box that appears, enter **Data Simulator** in the **Save the VI as:** box. Through this process, the original **Sine Wave** will be closed and returned to **YourName.llb**, while the new VI **Data Simulator** will opened, awaiting your editing.

By the way, this is the technique to use if you want to change the name of a VI that resides within a VI Library, a seemingly simple task, but one that can't be done directly. First, open the "incorrectly named" VI, then use **Save As…** to create a duplicate VI with the desired new name and store it in the VI Library. Finally, under the **File** menu, select **Edit VI Library…** and use this tool to delete the incorrectly named VI.

Modify the front panel of **Data Simulator** by using the Labeling Tool to rename the **dt** control as **Lower Limit**. Then add a new control **Upper Limit** and a new indicator **dx** to record the spacing of data points along the X-axis. It's probably a good idea to increase the **Digits of Precision** of this indicator from its default value of *2* to something like *6*.

As an aid in arranging these front-panel objects into a pleasing pattern, you might enjoy experimenting with the **Alignment Tool** and **Distribution Tool** in the toolbar. To use these tools, you must first use the Positioning Tool to highlight the group of objects on which you wish to operate.

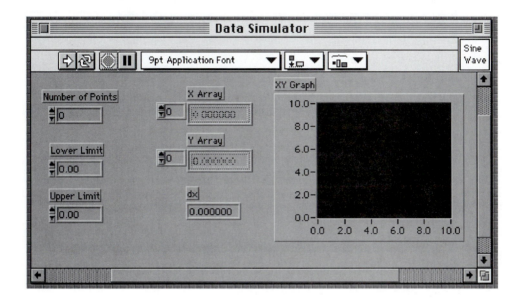

Pop up on the icon pane, select **Edit Icon**…, then design a new icon using the Icon Editor. Save your design by pressing the **OK** button.

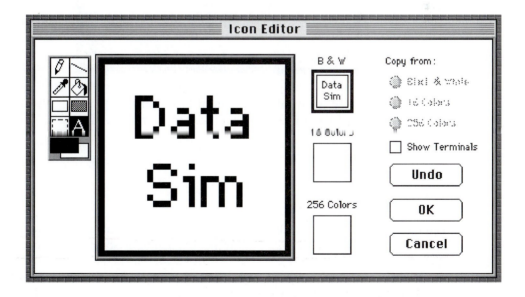

Pop up on the icon pane and select **Show Connector**. Pop up on the icon pane again and select a three-input, three-output connector from the **Patterns** palette.

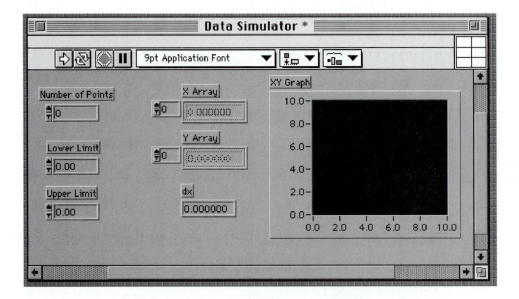

Now assign the six terminals according to the following scheme.

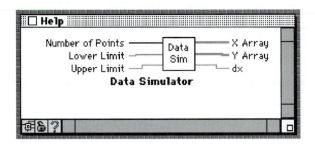

N equally spaced points X, ranging from X_0 (lower limit) to X_{N-1} (upper limit), can be created by first determining the necessary spacing between points Δx:

$$\Delta x = \frac{X_{N-1} - X_0}{N - 1}$$

Then

$$X_i = X_0 + i \, \Delta x \qquad i = 0, 1, 2, ..., N-1$$

Once the array of X-values is known, $Y_i = f(X_i)$ can be calculated by specifying the explicit form of the function f. Let's arbitrarily define f to be something simple such as $f(X) = 5 \, X^4$.

Switch to your block diagram and write code that produces the X and Y arrays in the manner described. One possible diagram is shown here.

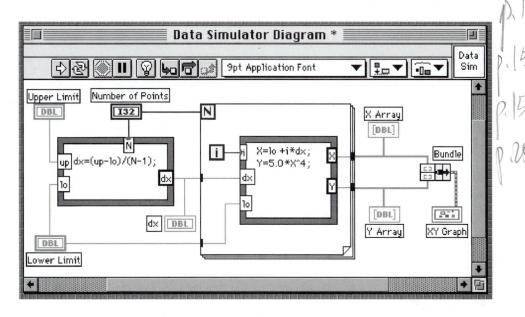

p. 181
p. 157
p. 159
p. 251

Switch back to the front panel. Input some values for the three controls (**Number of Points**, **Lower Limit**, **Upper Limit**), then run your program to verify that it works. Using **Save** in the **File** menu, store your completed program in **YourName.llb**, then close its window.

NUMERICAL INTEGRATION USING A SHIFT REGISTER

Here, we have a sequence of X_i values that are spaced apart by a constant step Δx. The solid curve represents a function $f(X)$ that has known values (shown as solid dots) at each X_i. Let $f(X_i) \equiv Y_i$.

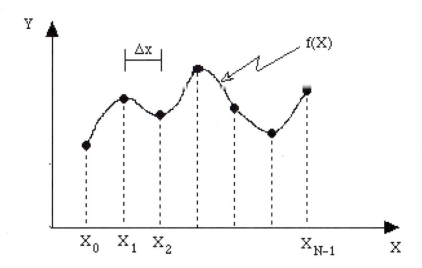

We wish to integrate the function f(X) between X_0 and X_{N-1}. This integral can be interpreted geometrically as the total area enclosed under the f(X) curve between these two end points. In the illustration, two adjacent dotted lines at the abscissas X_i and X_{i+1} define a columnar shaped area bounded on top by the curve f(X). We note that the sum of the areas of all such columnar shapes in the region between X_0 and X_{N-1} is equal to the integral we want to evaluate. Thus, if we have a general method for determining the area of each columnar shape, we can evaluate the integral.

Numerical Integration via the Trapezoidal Rule

The obstacle, of course, to writing a general formula for the area of a particular columnar shape is that each area has a curved top. In the *Trapezoidal Rule* for numerical integration, one assumes that the curve f(X) at the top of each columnar shape can be approximated by a straight line. This approximation becomes better and better as Δx is made smaller and smaller.

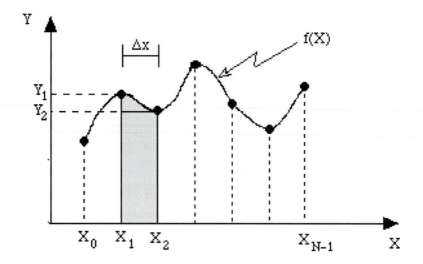

Then, for example, the columnar area between X_1 and X_2 is approximated as the trapezoidal area given by

$$Y_2 \Delta x + \frac{1}{2}(Y_1 - Y_2)\Delta x = \frac{1}{2}(Y_1 + Y_2)\Delta x$$

From this equation, we find that the Trapezoidal Rule is equivalent to assuming that the columnar shape defined by X_1 and X_2 is approximately a rectangle with height equal to the average of $f(X_1)$ and $f(X_2)$. So,

$$\int_{X_1}^{X_2} f(X)\, dX \approx \left[\frac{Y_1 + Y_2}{2} \right] \Delta x$$

where the graphical interpretation of this relation is shown next.

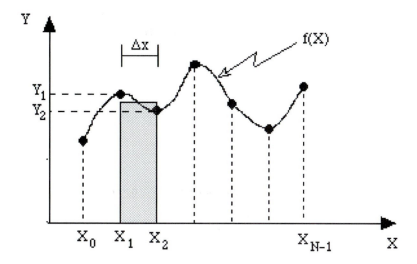

Using the Trapezoidal Rule, the approximation for the entire integral is then given in the following illustration.

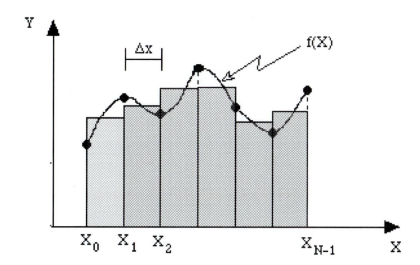

Thus

$$\int_{X_0}^{X_{N-1}} f(X)\, dx \approx \left[\frac{Y_0 + Y_1}{2}\right]\Delta x + \left[\frac{Y_1 + Y_2}{2}\right]\Delta x + \cdots + \left[\frac{Y_{N-2} + Y_{N-1}}{2}\right]\Delta x$$

or

$$\int_{X_0}^{X_{N-1}} f(X)\, dX \approx \left[\frac{1}{2}Y_0 + Y_1 + Y_2 + \cdots + Y_{N-2} + \frac{1}{2}Y_{N-1}\right]\Delta x \qquad (1)$$

In our program, we will find it most convenient to write this expression in the following equivalent way:

p. 147,

$$\int_{X_0}^{X_{N-1}} f(X) \, dX \approx \left[\{Y_0 + Y_1 + \cdots + Y_{N-1}\} - \frac{1}{2} \{Y_0 + Y_{N-1}\} \right] \Delta x \qquad (2)$$

On the right-hand side (RHS) of Equation (2), the first and second term in the square brackets are the sum of Y Array values and one-half the sum of the Y Array end points, respectively.

Trapezoidal Rule VI

Assume that an experiment has produced a set of N equally spaced (X,Y) data points, where the X-axis spacing between points is Δx and the variation of Y can be described by some function f(X). Let's write a VI that, given the Y Array and Δx, numerically integrates f(X) between the extremal X-axis values of X_0 and X_{N-1} using the Trapezoidal Rule.

Construct the front panel below and save this VI under the name **Trapezoidal Rule** in the **YourName.llb** Library. Provide plenty of **Digits of Precision** for the **dx** control and the **Value of Integral** indicator. Remember the **Array** input is formed by placing a **Digital Control** inside of an **Array** shell. Design an icon and assign the connector's terminals consistent with the Help Window shown.

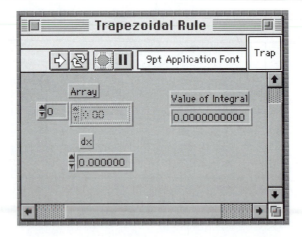

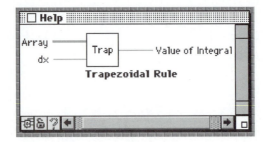

Now switch to the block diagram. Add a **For Loop** to your diagram and place a **shift register** on its boundary. To add the shift register, pop up on either of the For Loop's vertical borders and select **Add Shift Register**. You can then reposition the terminal pair using the Positioning Tool, if desired.

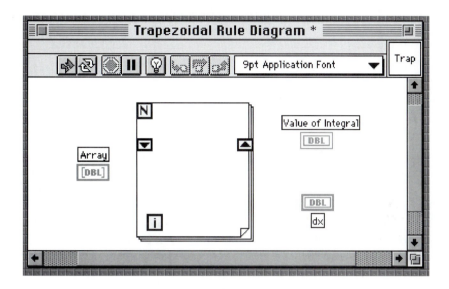

We are going to use the For Loop to compute the sum of the Y Array values of our data set. In a text-based language, we would write something along the lines of the following:

```
Sum = 0.0
For i = 0, N - 1
    Sum = Y(i) + Sum
```

To implement this code in LabVIEW, *initialize* the shift register to zero by wiring a **Numeric Constant** (defined as *0.0*) to its left terminal. Place an **Add** icon within the loop and wire an input and its output to the shift register's left and right terminal, respectively.

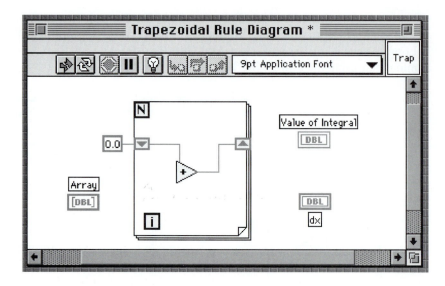

Now we take advantage of a handy LabVIEW convenience: In addition to its (previously studied) ability to build an array at a loop's output, auto-indexing can be used to sequence through an array at the loop's input. First, complete the following wiring, then I'll explain its meaning. Simply wire the **Array** control's terminal to the remaining **Add** input.

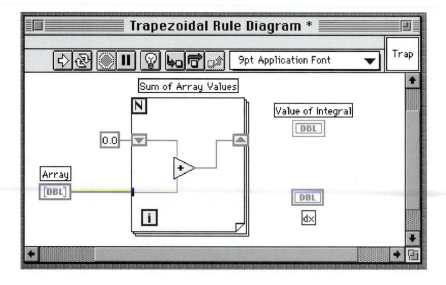

Note that the just-completed wire is thick (denoting an array) outside the loop, then becomes thin (denoting a scalar) inside the loop. Why is this so? Auto-indexing is activated in a For Loop by default, so upon execution, the loop will sequentially input one element from **Array** each time it iterates. That is, on the first loop iteration when

⊡ equals 0, the element of **Array** with index *0* will be input, on the second iteration when ⊡ equals 1, the element of **Array** with index *1* will be input, and so on until the end of the array is reached. As an added convenience, when auto-indexing is enabled on an N-element array entering a For Loop, LabVIEW automatically sets the loop's **count terminal** to N, thus eliminating the need to wire a value to 🅽.

The previous diagram works like this. The For Loop steps through the complete set of **Array**'s elements, accumulating the sum of these numeric values as it goes. During a particular iteration, the shift register's left terminal provides the sum accumulated up through the last iteration. The newly indexed array element is added to this sum and the result is stored in the shift register's right terminal for use in the next loop iteration. When the For Loop completes its operation, the sum of all of **Array**'s elements is contained in the shift register's right terminal. Thus, the value output from this terminal is the first term within the square brackets on the RHS of Equation (2).

We might flirt with the idea of neglecting the second term within the square brackets on the RHS of Equation (2). This term is an end-point correction to the integral expression. Since it involves the sum of just two array elements, it may be small compared to the first term calculated above. If we were to neglect this correction term, the integral would be determined by simply multiplying the For Loop output by Δx, as shown here.

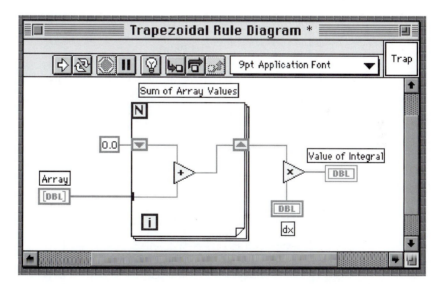

With not too much more work, though, we'll be able to write the code necessary to determine the end-point correction, which from Equation (2), is given by one-half the sum of the first and last Y Array elements. We can explicitly include this code on the present diagram, but let's take advantage of the modular nature of LabVIEW programming by tucking this code away within its own VI called **End Points**. You will then include **End Points** as a subVI on the **Trapezoidal Rule** diagram.

Open a new front panel and save it in **YourName.llb** under the name **End Points**. Place a **Digital Control** within an **Array** shell on the panel and label it **Array**.

Also label a **Digital Indicator** as **Half of End Point Sum**. Pop up on the icon pane, select **Edit Icon**…, then design an appropriate icon.

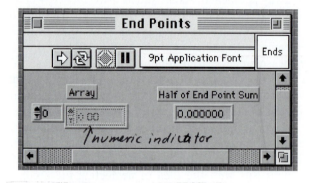

Pop up on the icon pane again and select **Show Connector**. Assign the terminals in a manner consistent with the following Help Window.

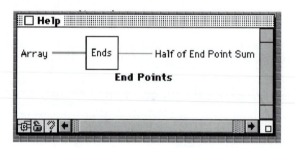

Switch to the block diagram. Here, given the N-element **Array**, we need to extract its first (index *0*) and last (index *N−1*) elements. The **Index Array** icon, found in **Functions>>Array**, performs this function. Its Help Window is shown below.

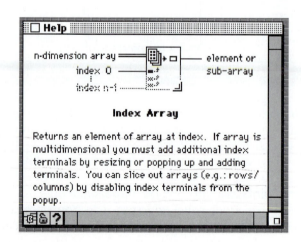

When operating on a two-dimension (or higher) array, proper configuration of this icon can require some thought. But for our present needs, involving the one-dimensional **Array**, **Index Array** is exceedingly easy to use. One simply wires the 1D array and an integer-containing **Numeric Constant** to the top and bottom inputs, respectively. The integer denotes the index of the array element to be extracted. Once extracted, the numeric value of this element appears at the icon's output terminal.

Another related icon that will prove useful is **Array Size**, also found in **Functions>>Array**. As shown in its Help Window below, this icon determines the number of elements present within an array. In the one-dimensional case, given an N-element 1D array at its input, the icon returns the integer (**I32**) N at its output.

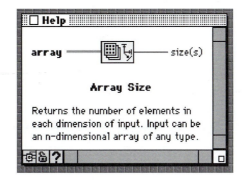

Complete the following block diagram. The left half of this diagram extracts the index *0* and index *N−1* elements from **Array**, while the right half adds these two numeric values together then divides by two. Save your work, then close **End Points**. The convenient **Decrement** icon is found in **Functions>>Numeric** and can be used to construct the integer *N−1*.

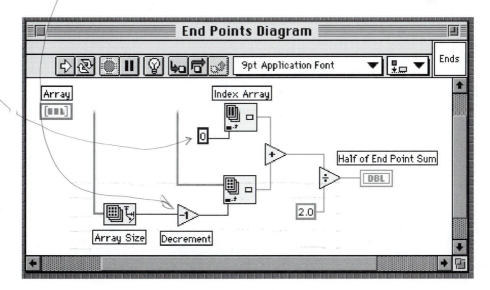

Alternately, with slightly more work, you can find $N-1$ using the **Subtract** icon.

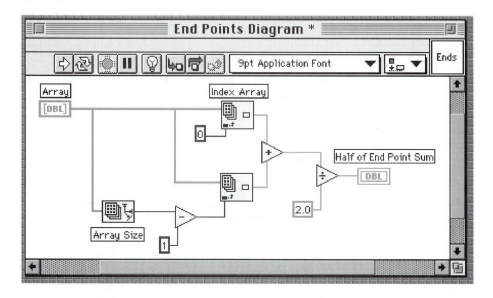

Now return to the **Trapezoidal Rule** block diagram. Include **End Points** as a subVI on this diagram in the manner shown below. The program now fully manifests all aspects of Equation (2).

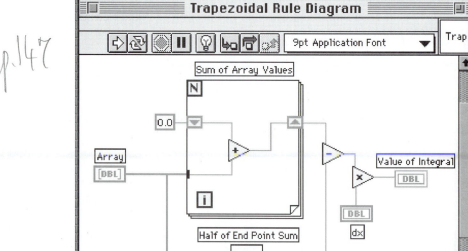

Let's see if the **Trapezoidal Rule** VI really works by supplying it with some known data from **Data Simulator**. Construct the following front panel and block diagram called **Trapezoidal Test**.

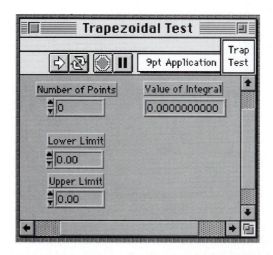

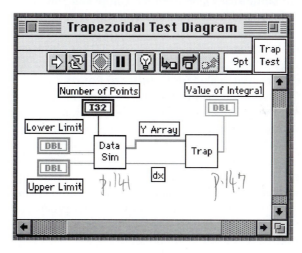

Assign the icon pane's connector terminals consistent with the following Help Window.

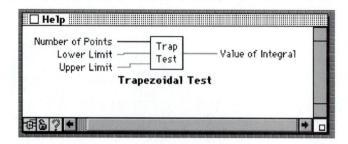

Use **Trapezoidal Test** to evaluate the integral $\int_0^1 5x^4 dx$ numerically. It is easy to show analytically that this integral is equal to exactly one. Knowing this true value, we'll be able to check the precision of our approximate numerical method.

Enter *0* and *1* on **Trapezoidal Test**'s front-panel **Lower Limit** and **Upper Limit** controls, respectively, and make sure that the Formula Node within **Data Simulator** is programmed to calculate $f(X) = 5X^4$. Since **Data Simulator** is a subVI within **Trapezoidal Test**, an easy way to open it is through use of the **Project** pull-down menu. Select **Data Simulator**'s icon from the **Unopened SubVIs** palette of the **Project** menu. **Data Simulator** will then open and you can perform the required check of its block diagram. After that, you can toggle back and forth between the front panel of **Trapezoidal Test** and the block diagram of **Data Simulator** using the **Windows** pull-down menu.

The appropriate value for **Number of Points** is more problematic. A larger value for this parameter makes Δx smaller. The smaller you make Δx, the less curvature the function $f(X)$ will have across the top of each columnar shape whose area is needed in the calculation of the integral. Thus, the less bent the top of the columnar shape, the more closely it resembles a straight line. So a large value for **Number of Points** will result in more accuracy for the Trapezoidal Rule approximation with the drawback that the program will take longer to run. Try running your program with a variety of choices for **Number of Points** and note the resulting **Value of Integral** in each case. Can you find a good compromise between high precision and acceptable run time? Remember, when evaluated analytically, the value of the integral is exactly 1.0000000....

Convergence Property of the Trapezoidal Rule

To explore the effect of **Number of Points** in determining the precision of the resulting integral value, write the following **Convergence Study** program. Place an **XY Graph** on the front panel. Using the Graph's pop-up menu, increase the Y-axis labeling **Digits of Precision** to *5* (from its default value of *1*) by selecting **Formatting**… under the **Y Scale** palette. This reformatting of the labels can also be done using the Graph's Palette. You'll probably have to resize the graph using the Positioning Tool to make the new labels completely visible.

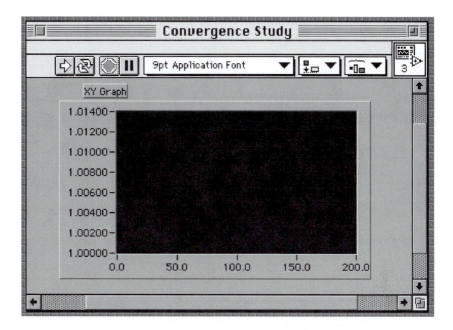

Next, write the following block diagram. This VI increments **Number of Points** from a small to large value, calculating (using **Trapezoidal Test**) and storing the integral value at each step, then displaying the resulting array of values on an **XY Graph**.

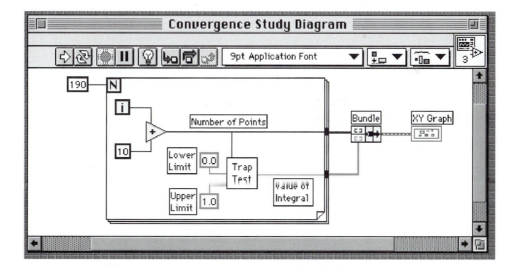

Run **Convergence Study**. You should find that a (diffusely defined) threshold value for **Number of Points** exists. Increasing **Number of Points** to values greater than this threshold value only produces incremental enhancement in the accuracy of the integral value. Such a study of the convergence properties of your calculational method would be quite useful in making an intelligent choice for the parameter **Number of Points**, if you were to use this VI to evaluate a complicated integral numerically.

NUMERICAL DIFFERENTIATION USING A
MULTIPLE-TERMINAL SHIFT REGISTER

In the last section, a shift register was used to recall a value created in the loop iteration immediately preceding the current one. In this section, we will explore how this register can be configured to remember values from not just one, but several previous loop iterations. We will explore this feature by writing a program that evaluates the derivative at each data point in a discretely sampled data set.

Assume we are given an equally spaced N-element set of (X,Y) data such that

$$X_i = X_0 + i \, \Delta x \qquad i = 0, 1, 2, \ldots, N-1$$
$$Y_i = f(X_i)$$

where

$$\Delta x = \frac{X_{N-1} - X_0}{N - 1}$$

and f is an analytic function. At the ith data point, the derivative can be evaluated numerically using the following expression that involves knowledge of the two neighboring data points:

$$\frac{dY_i}{dX} \approx \frac{Y_{i+1} - Y_{i-1}}{2 \, \Delta x} \qquad i = 1, 2, \ldots, N-2 \tag{3}$$

Since Y_{-1} and Y_N are unknown, this expression cannot be used at the array end points $i = 0$ and $i = N-1$.

Given a For Loop with the configuration of shift registers shown below, all of the quantities needed to calculate dY_i/dX via the prescription of Equation (3) will be available during the $(i + 1)^{th}$ iteration of the loop. Labels, which denote the indices of the array elements appearing in each shift register terminal during this particular loop iteration, are included in the illustration.

Thus, Equation (3) is coded by the following sub-diagram:

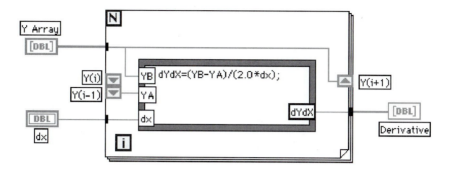

Initiating the execution of this diagram, however, presents us with a problem. Since Y_{-1} does not exist, the derivative of lowest index that can be calculated using Equation (3) is dY_1/dX, i.e., $i = 1$. To calculate this derivative during the For Loop's first iteration when $\boxed{i}$ equals zero, the top and bottom left terminals of the shift register must contain Y_1 and Y_0, respectively, while auto-indexing should place Y_2 at the Y Array input to the loop. To initialize the left terminals, we simply wire the desired values to them from outside the loop. In the case of the Y Array, we must somehow slice off the first two values so that the Y_2 value in the old array becomes the index-zero element in the new array. The new array then will have $N-2$ elements.

Let's write a VI called **Extract First Two** that manipulates the Y Array as desired above. Construct the following front panel, that accepts **Array** as an input and outputs its $i = 0$ and $i = 1$ elements to the **First Element** and **Second Element Digital Indicators**, respectively. In addition, **Extract First Two** outputs a new **Sliced-Off Array** that is formed by deleting the first and second element from the input **Array**.

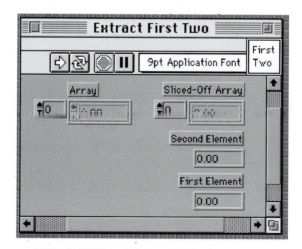

Design an icon for the icon pane and assign the connector terminals consistent with the following Help Window.

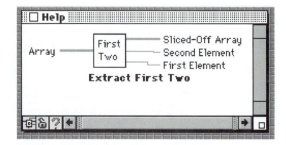

Then write the following block diagram.

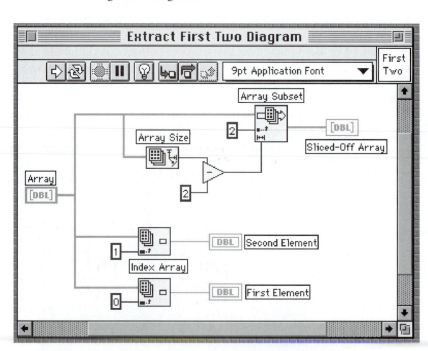

Here, we have used the **Array Subset** icon, found in **Functions>>Array**. Its Help Window is reproduced next. This VI receives an array as input and sets aside a given number (= **length**) of its sequential elements, starting at a prescribed index (= **index**). The VI then outputs this subset of elements as a new array. When using this icon, always keep in mind that arrays (as all other LabVIEW structures) are zero-indexed. That is, the first element has an index of zero, the second has index of one, and so on.

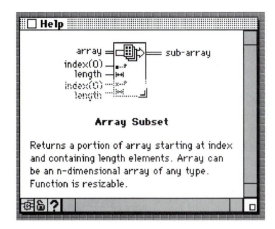

Now you are prepared to write **Derivative**. This VI will evaluate the derivative numerically at each point, except at the endpoints, of a given discretely sampled data set (supplied by **Data Simulator**). You will then display the calculated derivative on an **XY Graph** as well as in an Array Indicator. The front panel below contains all of the required input and output features. Construct it.

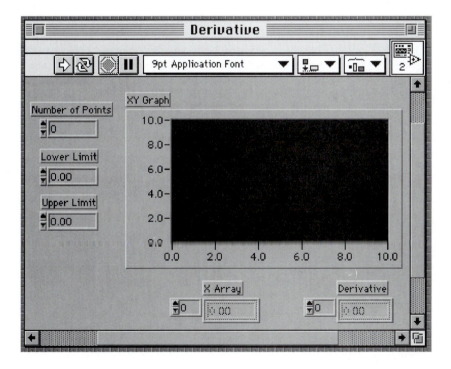

Complete the block diagram wiring shown here. This code will compute an array of $N-2$ elements whose first and last elements are dY_1/dX and dY_{N-2}/dX, respectively, where we are using the indexing appropriate to the original Y Array input. Some labels have been added to identify various wires.

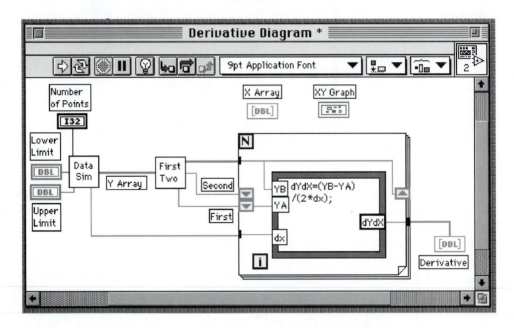

Finally, we need to slice the first and last elements off of the X Array output from **Data Simulator**, so that its indexing properly matches an X-value with the correct **Derivative** array element. Use **Array Subset** to perform this slicing operation, as shown. Then bundle the resulting X and Derivative arrays, and wire this cluster to the XY Graph terminal.

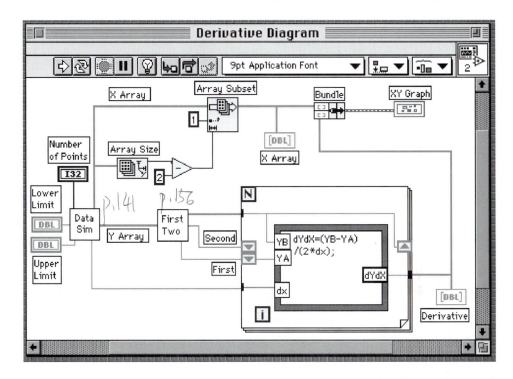

Return to the front panel and run your program. Take **Number of Points** to be *100* with the **Lower Limit** and **Upper Limit** as *0* and *1*, respectively. If you haven't made any changes, **Data Simulator** is still programmed with $f(X) = 5X^4$. Does **Derivative** produce a reasonable result? To judge if **Derivative** is operating correctly, it might be better to change the function within **Data Simulator** to something with a simple derivative such as $f(X) = 4X^2$.

To open **Data Simulator** so that a change in its block diagram can be made, you can use the **Project** and **Windows** menus as previously described. Here's another method for opening a subVI such as **Data Simulator**: Switch to **Derivative**'s block diagram, then double click on the **Data Simulator** icon. **Data Simulator**'s front panel will appear. Switch to its block diagram and change the equation within the formula node so that it calculates the function $f(X) = 4X^2$. Switch back to the **Data Simulator** front panel, close it, and save the changes when prompted. The **Data Simulator** block diagram will be automatically closed along with the front panel.

Run **Derivative**. Do you get the expected result, given that your input is now $f(X) = 4X^2$? Is the slope of the curve on your XY Graph as anticipated?

Try using your VI to calculate the derivative of $f(X) = 4X$.

GLOBAL VARIABLES

LabVIEW structures and subVIs present output data at the moment they complete execution. In some situations, you may require two independent VIs to pass information between each other, while each is still in operation. Or perhaps you might like a particular VI to select its present mode of operation based on recalling some aspect of its actions during a previous run time. You can accomplish either of these options by writing the desired information to a data file on your computer's hard drive and then subsequently reading it when needed. Such an approach, however, is extremely time consuming. Thankfully, a high-performance alternative—global variables—is available. Below, you will construct a global variable using a While Loop's shift register. Since this register resides within your computer's RAM, it is capable of storing and recalling information very quickly.

As an example of the need for a global variable, consider the following common situation. A LabVIEW program has been written to control the sequence of tasks necessary to implement a data-taking run. Your entire experiment then consists of repeatedly running this program to produce a series of runs, each taken under unique settings of the experimental parameters. However, to produce accurate results, some sort of calibration procedure must be performed. Assuming your instruments are stable, this calibration procedure need only be performed once. Thus, within your data-taking VI, you would like to activate the calibration procedure during the first data run, then, in the interest of saving time, ignore this procedure during subsequent runs. To write such a program, you require a memory element (which is the global variable) whose dedicated purpose is to recall whether the calibration procedure has been previously performed.

Let's write a global variable VI, then try to understand how it works. Construct the following front panel. The two **Vertical Switches** (found in **Controls>>Boolean**) input TRUE/FALSE values and the **Round LED** (also in found **Controls>>Boolean**) output a Boolean value. Label these objects as shown. Then design an icon in the icon pane. Save this VI under the name **Global Variable** in **YourName.llb**.

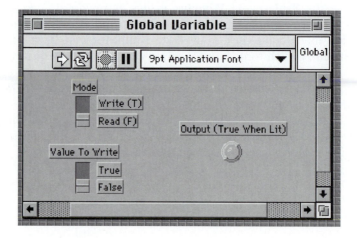

Finally, assign the connector terminals in a manner consistent with the following Help Window.

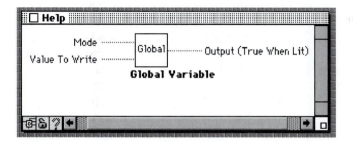

In writing the block diagram for **Global Variable**, you will use the **Select** icon (found in **Functions>>Comparison**). Its Help Window appears next. This icon is used to channel one of two possibilities to an output wire. The middle **s** ("select") input determines whether the **t** ("true") or **f** ("false") input will appear at the output. If a Boolean value of TRUE (FALSE) is connected to **s**, the value at the **t** (**f**) input is output.

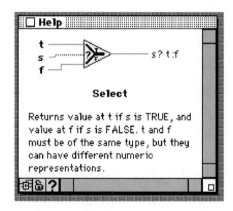

Now write the following block diagram as shown.

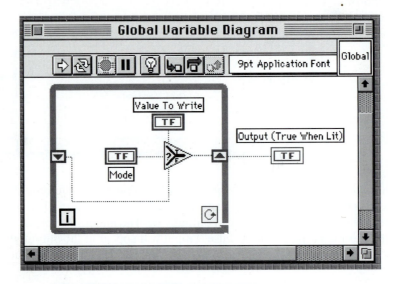

The operation of this diagram is determined by the lack of wiring to two of its objects. First, the While Loop's **conditional terminal** is unwired, so this terminal is always at its default value of FALSE. Thus, each time that the **Global Variable** program is called, the While Loop executes only a single iteration.

Second, nothing is wired from outside the loop to the shift register's left terminal. Thus the shift register is uninitialized and behaves as follows. The first time the VI is run after being loaded, the shift register contains its default value which is FALSE for a Boolean variable. (For numeric and string variables, the defaults are zero and empty, respectively.) From then on until the program is closed, each time the VI is run, the shift register contains the value left over from the last execution.

The above VI then functions as a global variable. If the **Mode** input is TRUE (Write), the value of the **Value To Write** control is written to the shift register. If the **Mode** input is FALSE (Read), the content of the shift register is recirculated. In either case, the content of the shift register is written to the **Output** indicator. Note that if you read the global-variable output before you write to it, you will read the default value for that data-type (e.g., FALSE for a Boolean variable).

Let's write a program called **Global Demo** that displays the operation of **Global Variable**. Write and save the following front panel containing a **String Indicator** found in **Controls>>String & Table**.

We are going to write a VI that prints out one text message (*"This is the first run"*) the first time it is run and another message (*"This is not the first run"*) each successive execution. Code the following diagram using your **Global Variable** icon and **String Constants**. Once written, here's how this diagram will function: First, the value of the global variable is read. Since nothing has been written to this Boolean global variable, the default value of FALSE will be output. This FALSE value is then passed to Select's **s** input, causing the **String Constant** containing *"This is the first run"* to appear at the **String Indicator**. Since nothing in the diagram affects the value of the global variable, this diagram will function in this exact manner during successive run times.

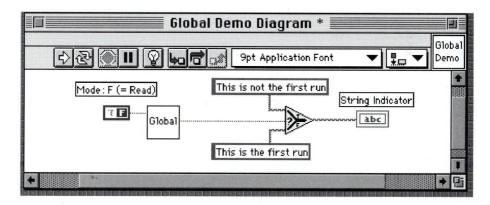

To complete the VI, we include some additional code that changes the value of the global variable to TRUE at the conclusion of the first run. In successive runs the string *"This is not the first run"* will be output by the **Select** icon. One method for coding this additional feature is shown in the next diagram. Here, the polymorphism of LabVIEW's built-in VIs, that is, the ability to operate on a variety of data-types (including scalar numbers, Booleans and strings as well as arrays of these various forms), is exploited in using the **Equal?** icon to check the equality of two character strings. At the end of the first run, **Equal?** will output TRUE, causing the TRUE Boolean constant to be written to the global variable. At the next run time, the TRUE global variable will select the *"This is not the first run"* string for output. Then the **Equal?** icon will generate a FALSE output, causing the global variable to be read benignly.

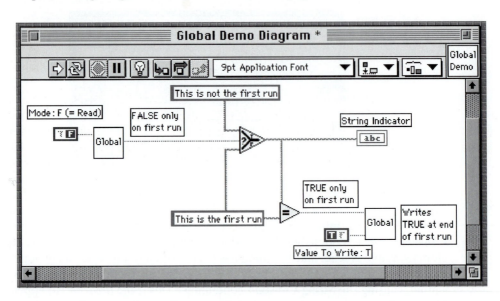

Close both of the **Global Variable** and **Global Demo** front panels. Now open the **Global Demo** VI and run it several times in succession. Close it, open it again, then run it repeatedly. Does it behave as expected? What happens if you duplicate this procedure while the **Global Variable** front panel is open?

The Case Structure

In a LabVIEW program, conditional branching is accomplished through the use of the *Case Structure*. This structure is analogous to an "if-then-else" statement in a text-based programming language. The Case Structure is found in **Functions>>Structures** and is Boolean by default. That is, by wiring a TRUE or FALSE Boolean value to its selector terminal ▣, the structure will execute either the code within its TRUE window or its FALSE window, respectively. Using the Positioning Tool, the selector terminal can be placed anywhere along the Case Structure's left border.

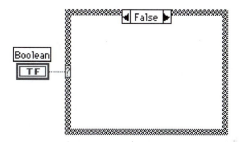

You may view only one of the Case Structure windows at a time. Above, the FALSE window is visible. To view the TRUE window, simply click the mouse cursor on the decrement (left) or increment (right) button in the ◀ False ▶ control at the top of the structure. The TRUE window will then appear as shown here.

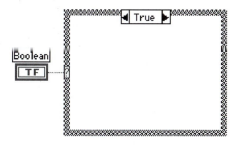

The Case Structure will automatically change its character from Boolean to numeric when you wire a numeric quantity to its selector terminal, such as the **I16** integer control shown next.

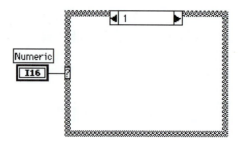

Initially only two Case windows (**0** and **1**) are available. Above, the 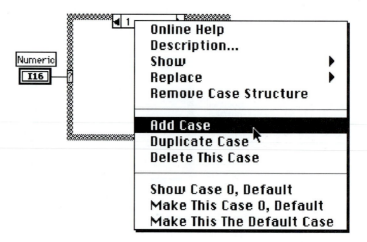 control indicates that the **Case 1** window is currently visible. You can easily add more cases by simply popping up on the Case Structure's border, then selecting **Add Case** in the resulting menu.

You will then find that the structure has three (**0**, **1**, and **2**) cases.

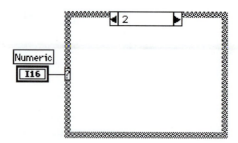

By repeating this procedure, you may add as many cases as you desire, unless you desire more than the maximally allowed $32768(=2^{15})$ cases! If you wire a floating-point number to the selector terminal, LabVIEW will round that number to the nearest integer value. If the number is negative or if the number is larger than the highest-numbered case, the case designated as *Default* will be selected.

Numerical Integration via Simpson's Rule

In this chapter, you will write a VI that numerically integrates a discretely sampled data set using *Simpson's Rule*. Assume that a curve $Y = f(X)$ has known values Y_1, Y_2, and Y_3 at X_1, X_2, and X_3, respectively, where the succession of three X values are equally spaced by a constant step Δx. In Simpson's Rule, one assumes that $f(X)$ in the region from X_1 to X_3 can be approximated by the quadratic $Y = A X^2 + B X + C$, as shown.

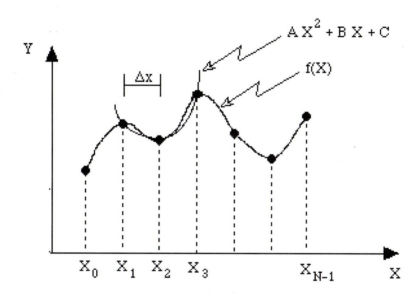

Evaluating the posited quadratic at the three known data points yields the following three equations involving the three unknown constants A, B, and C:

$$Y_1 = A X_1^2 + B X_1 + C$$
$$Y_2 = A X_2^2 + B X_2 + C$$
$$Y_3 = A X_3^2 + B X_3 + C$$

Using the fact that $X_2 = X_1 + \Delta x$ and $X_3 = X_1 + 2\Delta x$, linear algebra can be used to solve this set of equations for the three unknowns A, B, and C (try it yourself), yielding

$$A = \frac{1}{2(\Delta x)^2}[Y_1 - 2Y_2 + Y_3]$$

$$B = \frac{-1}{2(\Delta x)^2}[Y_1(X_2 + X_3) - 2Y_2(X_1 + X_3) + Y_3(X_1 + X_2)] \qquad (1)$$

$$C = \frac{1}{2(\Delta x)^2}[Y_1 X_2 X_3 - 2Y_2 X_1 X_3 + Y_3 X_1 X_2]$$

The area under the curve f(X) can then be approximated as follows:

$$\int_{X_1}^{X_3} f(x)dX \approx \int_{X_1}^{X_3} (AX^2 + BX + C)\, dX$$

$$= \frac{1}{3}A(X_3^3 - X_1^3) + \frac{1}{2}B(X_3^2 - X_1^2) + C(X_3 - X_1)$$

(2)

From Equation (1), we know the values of A, B, and C. Plugging these expressions into Equation (2), the right-hand side of (2) becomes

$$\frac{Y_1}{2(\Delta x)^2}\left\{\frac{1}{3}(X_3^3 - X_1^3) + \frac{1}{2}(X_2 + X_3)(X_3^2 - X_1^2) + X_2 X_3 (X_3 - X_1)\right\}$$

$$-\frac{Y_2}{(\Delta x)^2}\left\{\frac{1}{3}(X_3^3 - X_1^3) + \frac{1}{2}(X_1 + X_3)(X_3^2 - X_1^2) + X_1 X_3 (X_3 - X_1)\right\}$$

$$+\frac{Y_3}{2(\Delta x)^2}\left\{\frac{1}{3}(X_3^3 - X_1^3) + \frac{1}{2}(X_1 + X_2)(X_3^2 - X_1^2) + X_1 X_2 (X_3 - X_1)\right\}$$

Then, using $X_2 = X_1 + \Delta x$ and $X_3 = X_1 + 2\Delta x$, this expression greatly simplifies after some (brutal) algebra, yielding the following famous result known as the *Simpson's "Three-Point" Rule*

$$\int_{X_1}^{X_3} f(X)\, dX \approx \left[\frac{1}{3}Y_1 + \frac{4}{3}Y_2 + \frac{1}{3}Y_3\right]\Delta x$$

(3)

This relation provides an improved numerical integration method over the "two-point" Trapezoidal Rule. Note that the above formula gives the integral over an interval of size $2\Delta x$, so the coefficients add up to 2. Also note that, based on our derivation, we expect Simpson's "Three-Point" Rule will give exact results when the function f(X) is a polynomial of order 2 or less. Surprisingly, it can be shown that symmetries in the above equations lead to fortuitous cancellations, making Equation (3) exact even for polynomials of third order.

For a set of N data points then, where

$$X_i = X_0 + i\,\Delta x \qquad i = 0, 1, 2, \ldots, N-1$$
$$Y_i = f(X_i)$$

we can group the N X-values into a succession of "three-point" units, so that

$$\int_{X_0}^{X_{N-1}} f(X)\, dX \approx \left[\frac{1}{3}Y_0 + \frac{4}{3}Y_1 + \frac{1}{3}Y_2\right]\Delta x + \left[\frac{1}{3}Y_2 + \frac{4}{3}Y_3 + \frac{1}{3}Y_4\right]\Delta x$$

$$\cdots + \left[\frac{1}{3}Y_{N-3} + \frac{4}{3}Y_{N-2} + \frac{1}{3}Y_{N-1}\right]\Delta x$$

Let's call each term within square brackets a *"partial sum."* Then the above expression can be written as the following summation over partial sums

$$\int_{X_0}^{X_{N-1}} f(X) \, dX \approx \sum_i \left[\frac{1}{3} Y_{2i} + \frac{4}{3} Y_{2i+1} + \frac{1}{3} Y_{2i+2} \right] \Delta x$$

(4)

$$i = 0, 1, 2, \ldots, \frac{N-3}{2} \; (N \; odd) \; °$$

where the summation is over *(N − 1)/2* partial sums. We see then that this prescription for calculating the integral assumes that N is an odd integer, where N is the number of points in our data set. This, of course, is due to the fact that our N data points can only be successfully grouped into a sequence of three-point units if N is odd.

If given an even number of data points, one may calculate the integral numerically with the following approach: Use Simpson's Rule to find the contribution due to the first *N−1* (an odd integer) points, then the Trapezoidal Rule to calculate the contribution due to the region between the two last points.

$$\int_{X_0}^{X_{N-1}} f(X) \, dX \approx \sum_i \left[\frac{1}{3} Y_{2i} + \frac{4}{3} Y_{2i+1} + \frac{1}{3} Y_{2i+2} \right] \Delta x + \left[\frac{1}{2} Y_{N-2} + \frac{1}{2} Y_{N-1} \right] \Delta x$$

(5)

$$i = 0, 1, 2, \ldots, \frac{N-4}{2} \, (N \; even)$$

The first term on RHS of Equation (5) is the Simpson's Rule sum, which contains *(N − 2)/2* partial sums, and the second term is the Trapezoidal Rule applied to the last two elements of the Y-array.

ODD DETECTOR WITH ITERATION CALCULATOR USING A BOOLEAN CASE STRUCTURE

In applying Simpson's Rule to integrate a discretely sampled set of N data points, different equations must be used in the N odd and N even cases. We wish to write a program called **Simpson's Rule** that, given such an N-element data array, can determine whether N is odd or even and then implement the appropriate integration formula [Equation (4) or Equation (5)].

Let's first write a VI called **Odd?** which, given an integer N, will determine whether N is odd or even. **Odd?** will prove useful as a subVI in **Simpson's Rule**, where it will be used to determine the odd- or even-numbered nature of a given data set prior to integration. Since the number of partial sums in the entire Simpson's Rule summation is *(N − 1)/2* or *(N − 2)/2* if N is odd or even, respectively, we will include this calculation in **Odd?**.

Construct the following front panel with your own custom-designed icon. The Boolean indicator is a **Round LED** found in **Controls>>Boolean**.

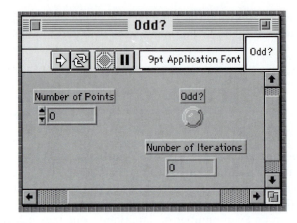

Assign the connector terminal according to the following scheme.

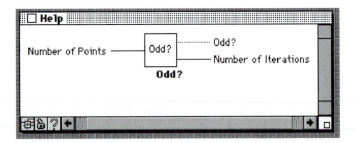

Switch to the block diagram and write the following code. The *mod(x,y)* function within the Formula Node, which corresponds to the modulo function of other programming languages, calculates the remainder when *x* is divided by *y* using the following algorithm: *remainder = x − y*floor(x/y)*, where *floor* truncates its argument to the next lowest integer. Thus *mod(N,2.0)* is *1* and *0* for N odd and even, respectively. In the diagram below, the Boolean indicator will be TRUE (FALSE) when **Number of Points** is odd (even).

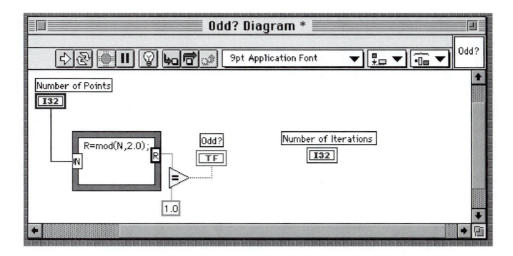

Return to the front panel and verify that the VI functions as expected by running it with several choices of **Number of Points**.

Now add a **Case Structure** (from **Functions>>Structures**) to your block diagram and wire the **Equal?** output to its selector terminal.

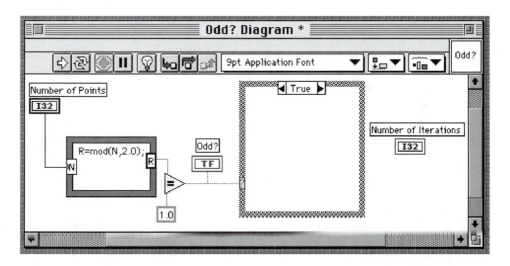

When N is even, the number of partial sums to be calculated is $(N - 2)/2$. Program this calculation into the Case Structure's FALSE window, then wire its output to the **Number of Iterations** indicator.

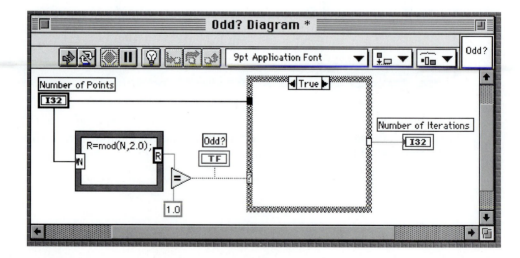

Note that the Case Structure's output tunnel is white and that the **Run** button is broken, indicating that the VI is not yet ready to run. Switch to the TRUE window and you'll find out why.

In a Case Structure, the data at all inputs (tunnels and the selector terminal) is available to all cases. Thus the **Number of Points** data tunnel, which was wired as an input in the FALSE window, will also make its data available in the TRUE window. However, a particular Case window is not required to use all available inputs. So even though nothing is connected yet, the input tunnel appears black, indicating a legal wiring configuration.

Output tunnels obey a different rule: If any one Case window supplies data to an output tunnel, all other windows must do so too. At present, the TRUE window lacks this requisite output-tunnel wiring and this unlawful situation is signaled by the white appearance of the tunnel. The broken Run button generically signals this diagram error.

When N is odd, the number of partial sums to be calculated is *(N − 1)/2*. Program this calculation into the Case Structure's TRUE window as shown next. As soon as you wire to the output tunnel, it will possess connections in all the Case windows (TRUE and FALSE). It will then turn black to denote it is now a legal output tunnel.

As an alternate method of coding the TRUE window, you might explore the following method, if interested. Code the FALSE window first. Then, since a very similar TRUE window is desired, pop up on the Case Structure's border and select **Duplicate Case**. A clone diagram will appear in the TRUE window, which you can then modify. A third window may appear in this process. You can delete it by using **Delete This Case** in the pop-up menu.

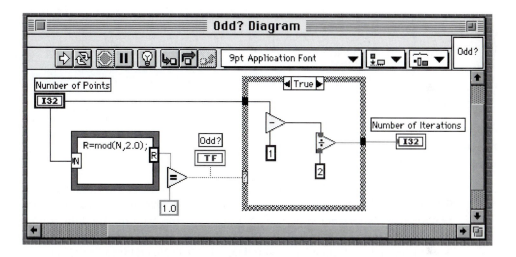

Run your program and verify that it functions properly. Then save the final version of this VI in **YourName.llb** as you close it.

Parenthetically, the Formula Node contains a conditional branching function that has the following syntax:

y = *(Condition)?Operation if Condition is TRUE:Operation if Condition is FALSE*;

Above, you wrote Case Structure-based code to accomplish the following logic:

```
If [mod(N, 2.0) = 1] then
    Number of Points = (N − 1)/2
else
    Number of Points = (N − 2)/2
endif
```

Such conditional branching could alternately be written using a Formula Node in the following way.

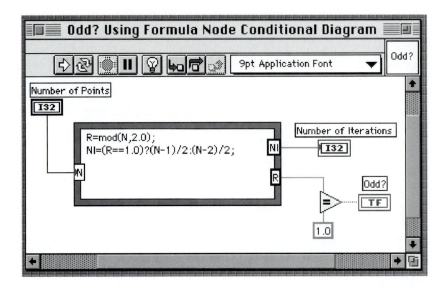

TRAPEZOIDAL RULE CONTRIBUTION USING A BOOLEAN CASE STRUCTURE

Another subVI needed for our numerical integration program calculates the second term on the RHS of Equation (5), that is, the Trapezoidal Rule contribution to the integral for an even-numbered data array. Let's write this VI now. Build the following front panel and design an icon in the icon pane. The Boolean control is a **Labelled Round Button**. Save the VI under the name **Even Ends** in **YourName.llb**. Assign the connector terminal in a manner consistent with the Help Window shown.

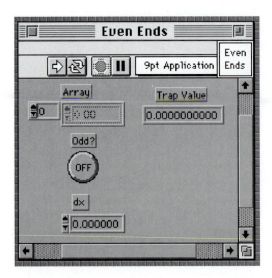

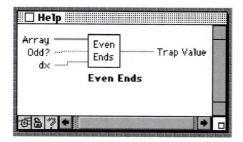

Now write the following diagram. If the array possesses an even number N of elements (**Odd?** is FALSE), the Case Structure's FALSE window determines the index of the second-to-last element ($i = N-2$), forms a sub-array consisting solely of the original array's last two elements ($i = N-2$ and $i = N-1$), then numerically integrates this sub-array using the **Trapezoidal Rule** VI.

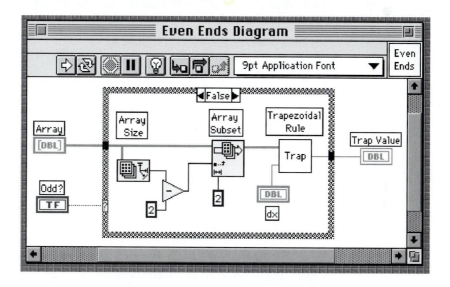

Code the TRUE window as follows, reflecting the fact that for odd-numbered arrays, the Trapezoidal Rule contribution is zero.

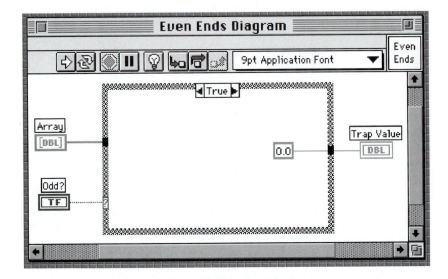

Be sure to save your final work on this VI as you close it.

SIMPSON'S RULE VI USING A NUMERIC CASE STRUCTURE

Now we are ready to write the **Simpson's Rule** VI. Assume that a set of N equally spaced (X,Y) data points exists, where the X-axis spacing between points is Δx and the variation of Y can be described by some function f(X). With the Y array and Δx as input, we wish to write a program that performs a Simpson's Rule integration of f(X) between the extremal X-axis values of X_0 and X_{N-1}.

Construct the following front panel. Using the **Format & Precision**… option in its pop-up menu, greatly increase the **dx** control's and **Value of Integral** indicator's **Digits of Precision** beyond their default values of *2*.

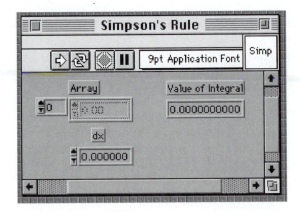

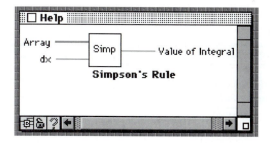

Now begin coding the block diagram as shown below. Here is our programming strategy: Use a pair of nested For Loops to implement Simpson's Rule. The outer For Loop successively picks out three-point units from the original input array using the **Array Subset** icon. The inner For Loop is used to calculate the partial sum due to each three-point unit. The summation of all the partial sums is accumulated in the outer For Loop's shift register; the number of For Loop iterations required to complete this summation is supplied by the **Odd?** icon. Since the entire input array must be passed to the **Array Subset** icon (as opposed to just one of its elements per loop iteration), you must pop up on the outer For Loop's tunnel and select **Disable Indexing**.

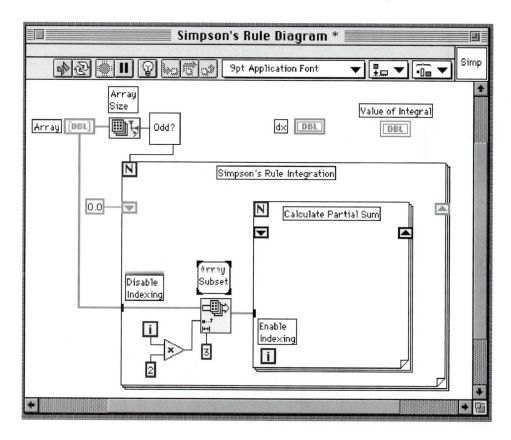

Now add the following code, which calculates each partial sum.

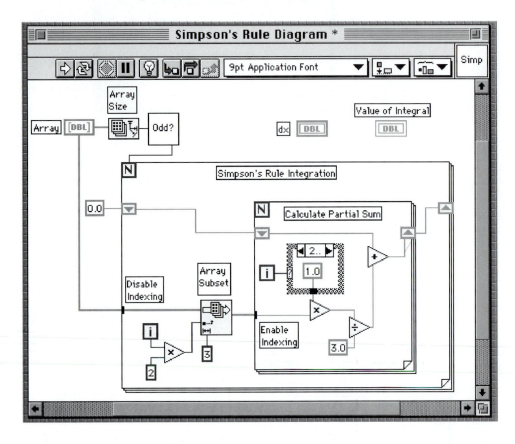

By enabling the indexing option on the inner For Loop, the three elements of the sub-array created by **Array Subset** will be passed into the loop one at a time. There is no need to wire the inner For Loop's **count terminal**; it will automatically be set to *3*. Within the loop, each of the three values is weighted by the proper factor consistent with Simpson's Rule. The inner loop's shift register accumulates the current partial sum. By initializing this register with the left terminal of the outer loop's register, the current partial sum is added to the accumulation of all past partial sums.

In the previous diagram, a numeric **Case Structure** is used to provide the correct 1–4–1 sequence of weighting factors dictated by Simpson's Three-Point Rule. Remembering that the iteration terminal begins counting at *0*, the three Case windows should appear as illustrated.

Complete the diagram to include the multiplicative factor of Δx and the Trapezoidal Rule end-point correction necessary for even-numbered arrays. Be sure to save your final work.

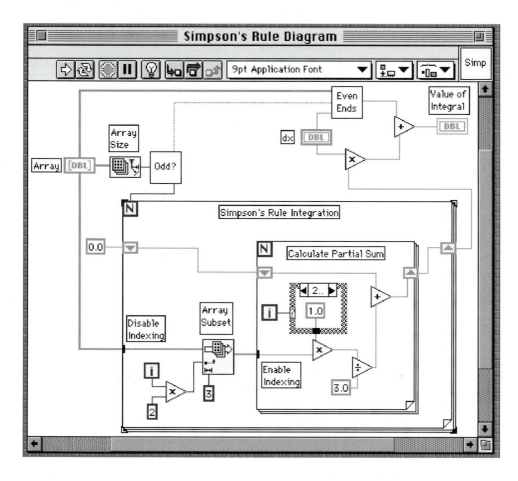

Let's see if the **Simpson's Rule** VI really works by supplying it with some known data from **Data Simulator**. Construct the following front panel and block diagram called **Simpson Test**

p. 180, 185

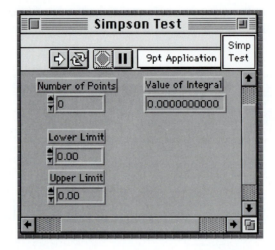

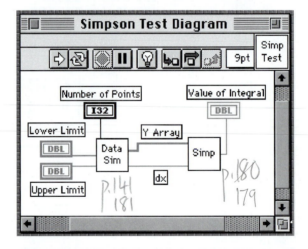

Use **Simpson's Test** to evaluate the integral $\int_0^1 5x^4 dx$ numerically by entering *0* and *1* on the front-panel **Lower Limit** and **Upper Limit** controls, respectively, and some appropriate value for **Number of Points**. Make sure that the Formula Node within **Data Simulator** is programmed to calculate $f(X) = 5\,X^4$. Try various values for **Number of Points** and see how this affects the precision of **Value of Integral**.

If you're interested, try using your VI to evaluate a harder integral. For example, the Riemann Zeta function is defined by

$$\zeta(s) = \frac{1}{\Gamma(s)} \int_0^\infty \frac{x^{s-1} dx}{e^x - 1}$$

This function is familiar to physicists because it appears in the theory of blackbody radiation, Bose-Einstein condensation, and the heat capacity of lattice vibrations. Let's take s = 3 and use the fact that $\Gamma(3) = 2! = 2$. Then

$$\zeta(3) = \frac{1}{2} \int_0^\infty \frac{x^2 \, dx}{e^x - 1}$$

Program **Data Simulator** to calculate the integrand. The Formula Node has a handy function *expm1(x)* that computes $(e^x - 1)$, so the following block diagram does the trick.

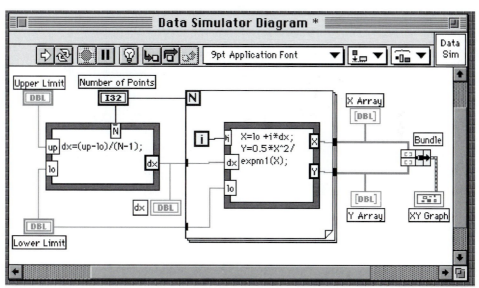

Run **Simpson Test** with appropriate choices for **Number of Points**, **Lower Limit** and **Upper Limit**. You'll encounter problems if you choose zero for **Lower Limit**. Find an acceptable solution to this problem. Also, you can't, of course, make **Upper Limit** equal to infinity. However, because of the nature of the integrand, it's acceptable to make **Upper Limit** something like *100*. Why? Optimize your choices to obtain the most accurate value possible for $\zeta(3)$. This integral cannot be done analytically. Nevertheless, high-precision numerical calculations (courtesy of Reed College's Richard Crandall) have determined that to a good (well, actually, very good!) approximation $\zeta(3) =$
1.20205690315959428539973816151144999076498629234049888179227155534183820578631309018645587360933525814619915779526071941841995998673283213776396837207900161453941782949360066719191575522242494243961563909664103291159095780965514651279918405105715255988015437109781102039827532566787603522336984941661811057014715778639499737523785277937030956025701853182790003076547107563048843320869711573742380793445031607625317714535444411831178182249718526357091824489987962035083357561720226033937858703281312678079900541773486911525370656237057440966221712902627320732361492242913040528555372341033077577798064242024304882815210009146026665382206962715520208227433500101529480119869019

17625951676366998171835575234880703719555742347294083595520886166620 2572853 7558130792825864872821737055661968989552662018776810629200817792338 13587682 8426412432431480282173674506720693507626895304345939375032966363775 75062473 3239923482883111957208761213609460551457394191100440365864268238615 06424055 4283127116707593518481 51488.

By comparison with this "correct" answer, how many decimal places of accuracy does **Simpson's Rule** deliver? Is this of satisfactory precision?

There are, of course, always many alternate ways to program a mathematical procedure. For example, with the following front panel,

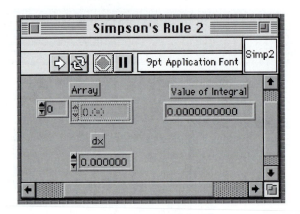

Simpson's Three-Point Rule can be implemented with the following diagram.

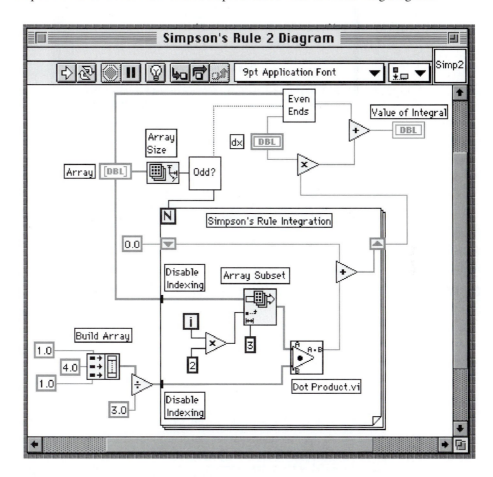

See if you can figure out how it works and, if interested, try coding and testing it yourself. The partial sum weighting is accomplished through the use of the **Dot Product.vi** icon found in **Functions>>Analysis>>Linear Algebra**. Its function is described in the Help Window found next. Also, you need to know that when the **Divide** icon has an array and a scalar numeric as inputs the icon's output is an array whose elements equal the input array's elements, each divided by the scalar.

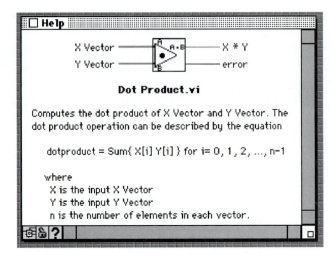

COMPARISON OF THE TRAPEZOIDAL RULE AND SIMPSON'S RULE

Finally, let's compare the convergence properties of the Trapezoidal Rule and Simpson's Rule methods and see if one proves to be the superior technique. To display the results of this study, place an **XY Graph** on a front panel and save it under the name **Convergence Study 2**. Enable autoscaling on the X- and Y-axes. Using its pop-up menu, hide the XY Graph's Palette, if desired.

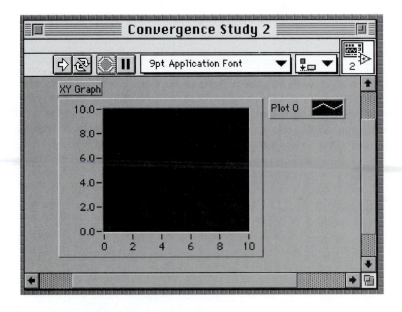

Now write the following block diagram that will numerically integrate an N-element array using both methods, where N varies from *10* to *200*, and records the results in arrays at the For Loop boundary. Here, I'm assuming **Data Simulator** is programmed to calculate the function $f(X) = 5.0\ X^4$. To plot the results from both methods on a single **XY Graph**, an XY cluster is formed for the Trapezoidal Rule as well as the Simpson's Rule result. A two-element array with the **Trapezoidal Rule** and the **Simpson's Rule** cluster as its first (index *0*) and second (index *1*) element, respectively, is built. Such an array of clusters is the required input for multiplots on an **XY Graph**.

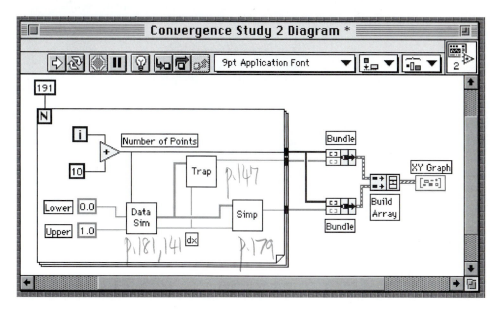

Return to the front panel. Using the Positioning Tool, resize the Legend downward to accomodate information for two plots.

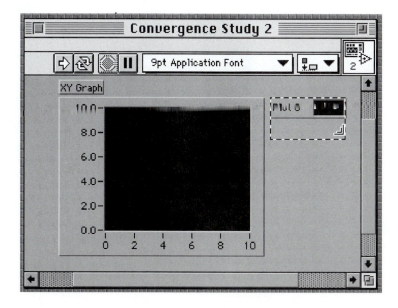

Using the Labeling Tool, highlight the text **Plot 0** in the Legend, then enter the text **Trap**.

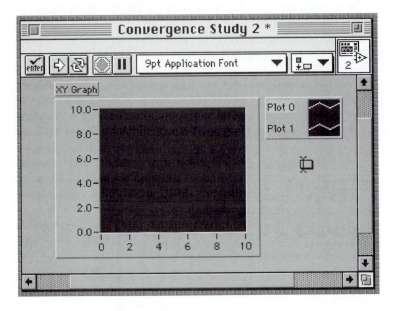

In a similar way, replace **Plot 1** by **Simp** in the Legend.

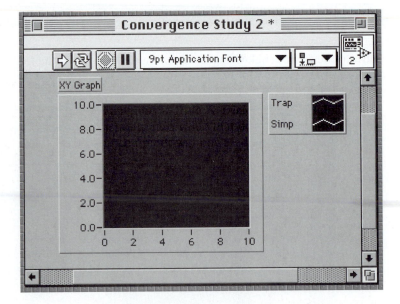

Now differentiate the plots by choosing unique plot characteristics in their pop-up menus. For example, below we choose the color red for the **Trap** plot, while the **Simp** plot remains the default color of white.

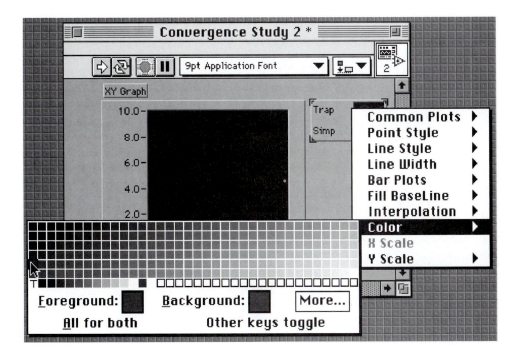

Finally, increase the **Digits of Precision** for Y-axis labeling from its default value of *1* to 5 by selecting **Formatting...** under the **Y Scale** palette of the XY Graph's pop-up menu. Alternately, you can make this adjustment using the XY Graph's Palette. Then resize the plotting region so that the new larger labels are fully visible.

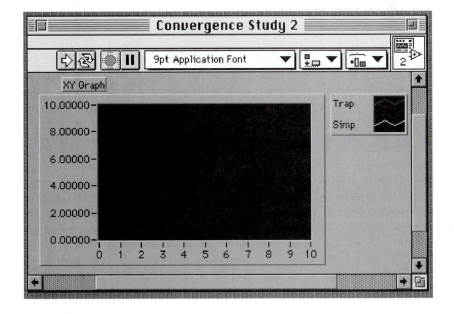

Run the VI. Does one of the methods converge to an accurate value for the integral more quickly (that is, with a smaller number of points) than the other?

Now that you understand the inner workings of these two numerical integration methods, check out LabVIEW's built-in icon **Numeric Integration.vi** found in **Functions>>Analysis>>Additional Numerical Methods**.

CHAPTER 7

The Sequence Structure

LabVIEW is a *dataflow* computer language. This approach to programming is contrary to the *control flow* mode in which most other programming languages operate. In control flow systems, the program elements execute one at a time in an order that is coded explicitly within the program. The series of sentence-like statements in, for example, C and BASIC programs, which describe the sequential execution of Procedure A followed by Procedure B followed by Procedure C, are manifestations of the inherent control flow fashion in which these text-based languages are structured. In contrast, LabVIEW program elements abide by the principle of dataflow execution. Obeying a condition termed *data dependency*, a given object on the block diagram (called a *node*) will begin execution at the moment *all* of its input data become available. After completing its internal operations, this node will then present processed results at its output terminals. The interesting advantage of dataflow programming is that several nodes (for example, A, B, and C) can, through LabVIEW's multitasking ability, effectively execute in parallel. Such parallel execution will occur whenever the receipt of node A's inputs overlaps in time with the execution sequence of node B and/or C. This multitasking ability has the potential to enhance system throughput, especially in situations (not uncommon to data-acquisition operations) where a portion of a particular node's execution time involves waiting. In LabVIEW, useful parallel processes can be executed while one node waits for an event, whereas in a control flow system such waiting periods are simply dead time, putting the entire execution sequence on hold.

Despite the impression that you might glean from the previous paragraph, LabVIEW programs can, if desired, be written with a guaranteed one-step-at-time sequence of node execution. A proficient LabVIEW programmer may exploit the data dependency maxim in coding a block diagram where the output of node A is used as a source of input data to node B. Such a diagram will mimic control-flow execution in that A must fully execute in order to enable the execution of B. The wiring configurations just described will many times provide an elegant solution to the need of ordered execution within a LabVIEW program. However, to create the correct configurations requires some skill on the programmer's part.

If an elegant data-dependent wiring configuration isn't forthcoming, LabVIEW provides the *Sequence Structure*, an explicit and foolproof way to obtain control flow within a program. The Sequence Structure, found in **Functions>>Structures**, looks like a single frame of movie film, as shown in the following illustration.

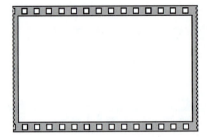

When it is initially placed on a block diagram, the Sequence Structure contains only one (unlabeled) frame. You can add any number of new frames, though, by popping up somewhere on the structure's border and selecting **Add Frame After**.

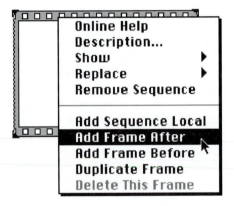

In the above example, the original and the newly created frame will be labeled **0** and **1**, respectively, and the Sequence Structure will now appear as

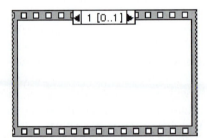

Note that, although the Sequence Structure may consist of several frames, only one frame is visible at a time. The ◄ 1 [0..1] ► control indicates that the **frame 1** window is currently visible; the list in square brackets informs you of the total span of frames programmed into the structure (in this example, there is **frame 0** through **frame 1**). Just as we found for the Case Structure, you may view other frames by clicking on the arrows in this control.

A Sequence Structure executes **frame 0**, followed by **frame 1**, then **frame 2**, and so on until the last frame executes. Thus, by placing objects within different frames, you can use the Sequence Structure to control the order of execution among nodes that are not naturally coupled through data dependency. In addition, data can be input to the structure through tunnels and the data at a particular input tunnel is available to all frames. Each frame can also emit output data through a tunnel, but consistent with dataflow principles, this data will not be output from the structure until its last frame completes execution.

Local variables may be defined within a Sequence Structure, allowing you to pass data from one frame to any subsequent frame. To create one of these variables (sensibly called a *sequence local*), simply pop up on the structure's border and choose **Add Sequence Local**. The sequence local terminal will appear at the structure's border as a small yellow box, which you may then drag to any unoccupied location on the border.

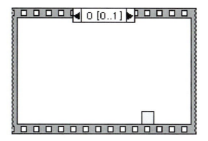

Once you wire a source of data to the sequence local within a particular frame, an outward-pointing arrow appears in the terminal of that frame. The sequence locals in all subsequent frames then will contain an inward-pointing arrow, indicating that the terminals may be used as a data source in each of their respective frames.

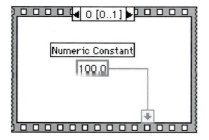
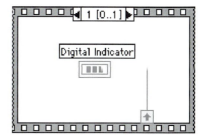

EVENT TIMER USING A SEQUENCE STRUCTURE

p.199)

A timer, which measures the duration of a given event, provides an example of an instrument in which operations must proceed in a specific order—a START initiates the timing action, followed by a STOP which terminates it. In building a LabVIEW timer VI then it is perfectly natural to employ a Sequence Structure on the block diagram. Let's build such a timer and use it to measure something interesting. We will use our timer to compare the time it takes each of LabVIEW's two repetitive-operation structures, the For Loop and the While Loop, to build a given array of data. Our findings may surprise you!

Build the following front panel and save it in **YourName.llb** under the name **Loop Timer-Sequence**. Change the data-type of the **Number of Points** control and the **For Loop Array** and **While Loop Array** indicators to **I32** and then resize these objects so that they may accommodate larger integer values. Change the precision of the **For Loop Time (sec)** and **While Loop Time (sec)** indicators from the default value of *2* to *3*.

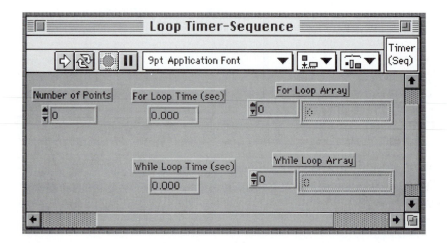

Switch to the block diagram. The basis of our timer will be the **Tick Count (ms)** icon. This VI is found in **Functions>>Time & Dialog** and its Help Window is reproduced next. **Tick Count (ms)** outputs a value that denotes the number of milliseconds that have elapsed since your computer was powered on.

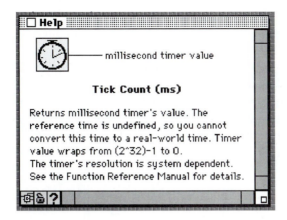

Our strategy for measuring elapsed time will be this: Call **Tick Count (ms)** at the START of an event and call it at the event's STOP. Then by subtracting the two **Tick Count (ms)** output values, the event's elapsed time will be determined.

Place a **Sequence Structure** on the block diagram. Inside the initial frame, place a **sequence local** on the border (by popping up on the border and selecting **Add Sequence Local**), then wire a **Tick Count (ms)** icon to it. This sequence local will contain your computer's internal clock value at the START of the For Loop, which will be contained in the next frame.

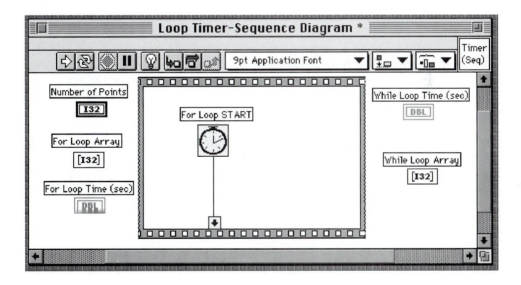

Create **frame 1** by selecting **Add Frame After** from the Sequence Structure's pop-up menu. Inside **frame 1**, code a For Loop that creates an N-element array, where each array element has a numerical value equal to the value of its index. Output this array to the **For Loop Array** indicator on the front panel so that you may view its elements. N is determined by the **Number of Points** control that is wired from outside the Sequence Structure, through a tunnel, to the For Loop's **count terminal**.

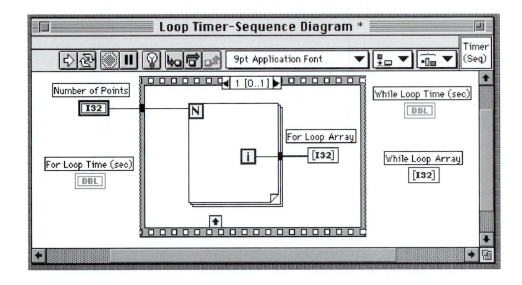

Create **frame 2**. Place a **Tick Count (ms)** icon there to ascertain your computer's clock value at the For Loop STOP. Then use **Subtract** to find the difference between the For Loop STOP and START (stored in the sequence local) time in milliseconds. Divide this time by *1000* to convert it to a more readable value in seconds and output the result to the front panel's **For Loop Time (sec)** indicator. Finally, the **Tick Count (ms)** output in this frame will also be used as the While Loop START value. Create a new **sequence local** and wire this icon to it, so that its output will be available for use in later frames.

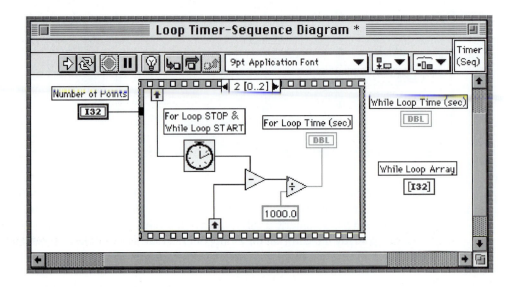

Create **frame 3** and enclose within it the While Loop-based code shown. As we learned in Chapter 2, by wiring $N-1$ to the lower terminal of the **Less?** icon, the While Loop will create an N-element array. Remember to **Enable Indexing** at the tunnel; the While Loop default is **Disable Indexing**.

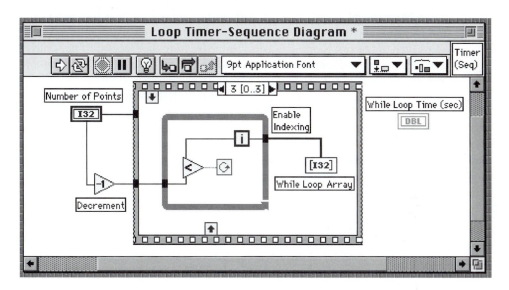

Finally, create **frame 4**. Here, determine the While Loop STOP through the use of a **Tick Count (ms)**, then subtract the While Loop START from it. Divide by *1000* to convert to seconds and output this value to the front panel's **While Loop Time (sec)** indicator.

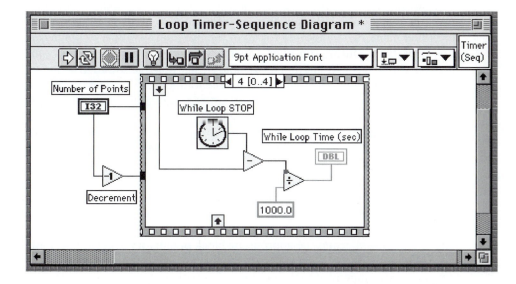

The VI is now complete, so return to the front panel and save your work. We will use **Loop Timer-Sequence** to place the For Loop and While Loop in head-to-head competition to see if either provides superior performance in the task of assembling an N-element array.

Prior to running this program, you will need to input an appropriate value for N in the **Number of Points** control. The proper choice for N will depend on the speed and memory capabilities of your computer. You should make N large enough that it takes each of the two loops at least a sizable fraction of a second (if possible, more than 0.10 seconds) to complete the array-building process. For times less than a few tenths of a second, the accuracy of **Tick Count (ms)** becomes an issue, as discussed later. However, if you make N too large, you will find that **Loop Timer-Sequence** produces a run-time error indicating that the program has overwhelmed your computer's free memory. You can possibly circumvent this problem by allocating more memory, if available, to LabVIEW.

To get started, choose a value for N in the range of *1,000* to *1,000,000*. Then, through an iterative process, determine an appropriate N value for your system.

Once you've established a good value for N, get a fresh start by closing **Loop Timer-Sequence**, then reopening it. Input your N value, then run the VI. Note the resultant values of **For Loop Time (sec)** and **While Loop Time (sec)**. You should find that the For Loop produces the array significantly faster than the While Loop does. Can you speculate why? You may wish to peruse the Array Indicators to verify that the exact same array is being produced by the For and While Loop methods.

Without closing the VI, try rerunning it a second, third, fourth, and more times and noting **For Loop Time (sec)** and **While Loop Time (sec)** for each run. You should find that the second run produces output times noticeably shorter than the initial program execution (that is, the first run after opening the VI). Subsequent runs then reproduce these same times, with only a small fluctuation on the order of 1 to 55 milliseconds (depending on your particular computing system).

To understand these observations, you must first understand how LabVIEW uses memory. When LabVIEW launches, a single block of memory (either physical RAM or virtual memory) is allocated for all of the program's editing, compiling, and execution operations. During run time, a memory manager allocates memory for tasks as needed, with the constraint that arrays and strings must be stored in contiguous blocks of memory. If the memory manager is unable to find a block of unused memory that is large enough for a particular string or array, a dialog box appears to indicate that LabVIEW was not able to allocate the required memory. On a Macintosh, such a situation can possibly be remedied by increasing the memory allocated to LabVIEW using the **Get Info** dialog in the **Finder**. You'll need to restart LabVIEW so that the new memory allocation can take place.

Now let's consider what happens when the auto-indexing feature of a looping structure is used to build a data array. In the case of a For Loop, LabVIEW can predetermine the size of the array to be built based on the value wired to the Loop's count terminal. Thus the memory manager is only called once as the loop initiates execution and at that time allocates the appropriate block of memory necessary to hold the array.

With a While Loop, however, the final array size cannot be known in advance. A new element is appended to the existing array with each loop iteration and this process continues until a FALSE Boolean value at the conditional terminal causes the loop to complete its execution. Because the array is constantly increasing in size as the loop iterates, the memory manager must be called continually in order to find an appropriately sized chunk of RAM to hold the ever-growing data array. These repeated calls to the manager take time, especially if memory starts to become scarce. In a memory-tight situation, the manager may take extra time as it tries to shuffle around other blocks until a suitable space opens up.

Thankfully, because of some built-in LabVIEW intelligence, a While Loop's array-building capability is not as handicapped by the memory manager as it might appear from the above description. To avoid calling the manager with each iteration, the While Loop's auto-indexing feature instructs the manager to allot enough new memory to store not just a single new array element, but instead a large number of additional array elements each time it is called. Through this trick, the number of memory-manager calls necessary in building a given sized array becomes rather small. Of course, when the loop completes its execution, there most likely will be some unused memory associated with the array because of the over-generous manner in which the manager delivered memory on its last call. Thus, when the loop terminates, LabVIEW simply directs the manager to dissociate this excess memory from the now-complete array. The result of this shrewd use of the memory manager is that the While and For Loops' array-building capabilities are not widely divergent (say, by powers of ten) in their performance, as you discovered through **Loop Timer-Sequence**. The slower speed of the While Loop execution is partly due to the comparison operation necessary to determine if further iterations are required. You can verify this point by adding a **Less?** icon within the For Loop as shown. How do **For Loop Time (sec)** and **While Loop Time (sec)** compare once this addition is made to the VI?

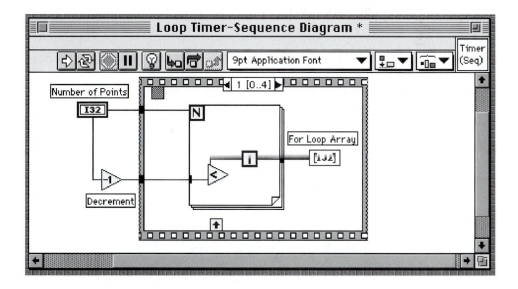

Next, we noted that the **Loop Timer-Sequence** VI runs slower during its first run than it does during its subsequent executions. This observation can be traced to use of the memory manager. During the program's first run, the manager works out the memory allotments peculiar to the particular need of that VI. During subsequent runs, many of the first-run memory assignments are simply reused, diminishing the time-consuming use of the manager.

Finally, what causes the small fluctuations in execution times during the second, third, fourth,... runs? These fluctuations are caused by the intrinsic accuracy of LabVIEW's built-in timing VIs. These icons have names such as **Wait (ms)** and **Tick Count (ms)**, leading one to suspect that they have a precision on the order of one millisecond. This conjecture, however, is in fact the reality only on the latest-model computing systems. These icons mark time by counting interupts to the system's CPU which, on the Power Macintosh and on Windows 95/98/NT machines (except those running with slower processors such as a 80386), do indeed occur at 1 millisecond intervals. However, for 68K Macs and Windows 3.1 systems, the interrupt interval is 17 ms and 55 ms, respectively. Thus the resolution you obtain from **Wait (ms)** and **Tick Count (ms)** depends on the details of your particular system and, in the best case, is 1 millisecond. It is this inherent inaccuracy of **Tick Count (ms)**, when executing on your particular system, that accounts for the observed timing fluctuations. The lesson to be learned here is that LabVIEW's timing VIs provide a simple and effective method of measuring the duration of an event, provided that millisecond resolution is adequate. On the other hand, in situations that demand sub-millisecond timing (such as is common in acquiring a sequence of analog-to-digital conversions), these icons are totally inadequate. In later chapters, we will find that the microsecond-resolution clock on a National Instrument's data acquisition (DAQ) board can be used in high-accuracy timing applications.

If interested, you might try exploring the performance of alternate array-building diagrams. These diagrams purposely avoid use of a loop's auto-indexing feature in an effort to make explicit what this feature does automatically. For example, try the following For Loop-based method, which initializes an appropriate-sized array (with each element defined as zero) prior to the loop structure, then avoids further calls to the memory manager through the use of **Replace Array Element** within the loop itself. Use Help Windows to understand the functioning of the unfamiliar icons.

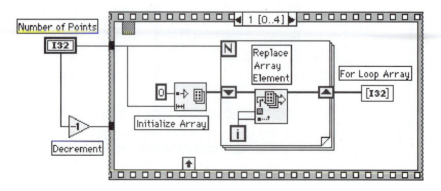

A While Loop-based diagram is shown next, which initializes the shift register with an empty array, then appends elements to it with each loop iteration. You will need to pop up on the top input of **Build Array** and select **Change to Array**. Why is this diagram such a poor performer? In non-demanding applications (e.g., involving small-sized arrays), however, this diagram is commonly used to perform "real-time" graphing (with a calibrated X-axis) inside a While Loop by feeding the **Build Array** output to the terminal of a **Waveform Graph** (see the last pages of Chapter 12).

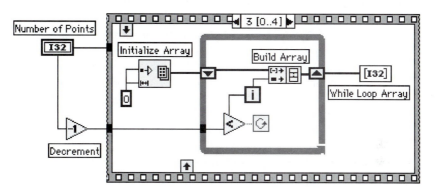

Another While Loop-based diagram initializes a larger-than-needed array (a 100,000-element array is shown; your choice of N may require something different), sequentially places data values in the first N elements of this array through the use of **Replace Array Element**, then lops off the excess remainder of the array using **Array Subset**. Why does this diagram execute quickly?

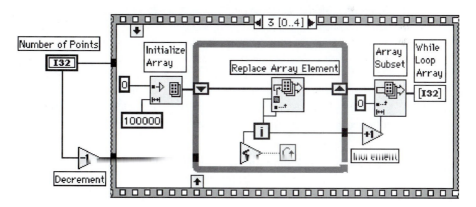

EVENT TIMER USING DATA DEPENDENCY   p. 192,

A skillful programmer can exploit LabVIEW's data dependent mode of execution to force two or more nodes to execute sequentially. Thus it is almost always possible to program a diagram without the use of a Sequence Structure.

In some cases, it is natural for node A's output to be used as the input of node B. Then these "chained together" nodes will execute in the order A-B. LabVIEW's File I/O icons provide such an example because all of these VIs have an input and output called **file path** and **new file path**, respectively. Let A and B be two file-related icons such as **Write Characters To File.vi** or **Write To Spreadsheet File.vi**. If A is (somehow) provided an input **file path**, when it completes its execution, **new file path** is output. By wiring this output to B's **file path** input, one guarantees that A executes fully before B. Look back at your work in Chapter 4. We implemented this scheme in the **Spreadsheet Storage** VI. In later chapters, you will find that most of the data acquisition-related icons have **error in** input and **error out** output terminals that can be used to chain together several related nodes to ensure sequential execution.

Many times, however, the output of A does not constitute a natural input for B. In such situations, ordered execution can still be obtained by an alternate method called *artificial data dependency*. In this scheme, it is the mere arrival of data, with no regard to its actual value, that triggers the execution of a node. An example of artificial data dependency will be constructed on the following pages.

Let's try to write a timer diagram without the use of a Sequence Structure. First, open **Loop Timer-Sequence** and use **Save As...** to create a new VI called **Loop Timer-Data Dependency** in **YourName.llb**. Keep the front panel unchanged as shown.

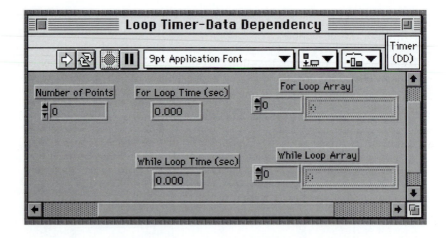

Switch to the block diagram. Delete the **Sequence Structure**. Code the For Loop subdiagram shown next, which creates an N-element array with the numerical value of each element equal to its index. Place a **Tick Count (ms)** icon on the diagram and wire its output to the border of the For Loop. **Tick Count (ms)**'s output then is an input to the **For Loop** meaning that the loop cannot initiate execution until **Tick Count (ms)** obtains a value. This clock value will be used as the For Loop START. The START value is never used within the For Loop, so it is simply the arrival of Tick Count's output that triggers the loop execution—an example of artificial data dependency.

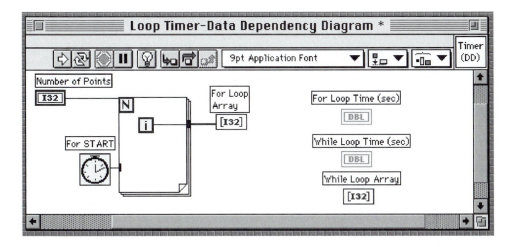

After the For Loop ceases execution, we want to obtain the For Loop STOP value using a **Tick Count (ms)** icon, then calculate the elapsed execution time in seconds. The trick for assuring this sequence of operations is to enclose the "STOP-value" code within a While Loop, as shown below. The **conditional terminal** of this While Loop is left unwired, meaning the loop will iterate only once. This single iteration occurs upon receipt of a value at the loop's input tunnel. Wire the START value, through the For Loop, to an output tunnel (disable the loop's auto-indexing feature). Then wire this For Loop output over to the While Loop input. The completion of the For Loop will then trigger the execution of the While Loop iteration, as desired.

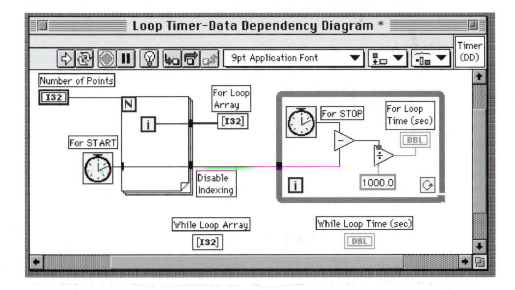

Now, wire the **Tick Count (ms)** output to an output tunnel on the While Loop's border as shown in the following illustration. We will soon code a While Loop, which creates an N-element array, and this output tunnel will contain the While Loop START value.

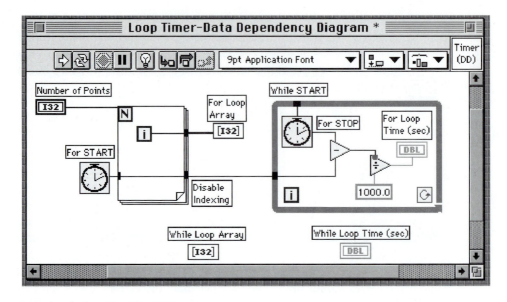

Construct the array-building While Loop shown below. You will have to enable auto-indexing at the loop border. Use artificial data dependency to trigger its execution with the While Loop START value.

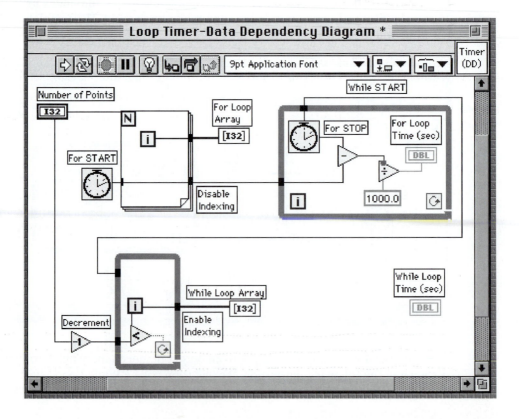

Complete the diagram as follows to calculate the time it takes for While Loop-based array building. Save your work.

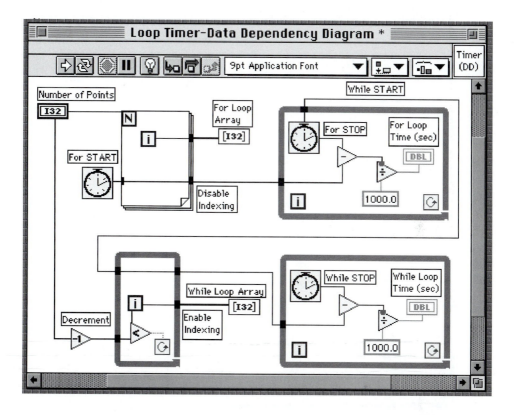

Return to the front panel and run the VI with your optimized choice for **Number of Points**. Check to see that the expected N-element array is built by both the For Loop and the While Loop. If your diagram is correct, the performance of **Loop Timer-Data Dependency** should equal that of **Loop Timer-Sequence**.

To cement your understanding of the manner in which this VI executes, let's implement one of LabVIEW's handiest (and coolest!) debugging tools called *Highlight Execution*. Input something small, such as *10*, for **Number of Points** on the front panel, then switch to the block diagram. In the toolbar, enable **Highlight Execution** by clicking on the button containing the light bulb.

Then click on the **Run** button. The VI will execute in slow-motion animation, with the passage of data marked by bubbles moving along the wires and important data values given in automatic pop-up probes. Because the diagram executes in slow motion, the values for **For Loop Time (sec)** and **While Loop Time (sec)** obtained under Highlight

Execution will be meaningless. However, this mode of execution provides a beautiful visual demonstration of the concept of artificial data dependency.

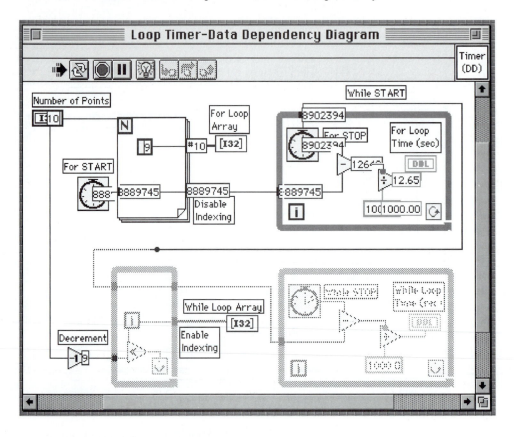

Use **Highlight Execution** liberally in your future work. This LabVIEW feature is an invaluable debugging tool for troubleshooting VIs that are causing you problems.

Built-In Analysis VIs—Curve Fitting

Before You Come To Lab

In the coming lab exercises you will be using a thermistor, which is a device with a temperature-dependent resistance, as a temperature sensor. Thermistors have the nice feature that their resistance versus temperature (R-T) curve can be fit to a well-known analytic function, facilitating their use as a calibrated thermometer. In the following table, the R-T data for a particular thermistor (Radio Shack Model #271-110) is given. If you will be using a different thermistor in the experiments described in Chapters 10 and 11, please use the R-T data for the thermistor that will be used in your particular experimental set-up.

Temperature in Celsius	Kilohms
−50	329.2
−45	247.5
−40	188.4
−35	144.0
−30	111.3
−25	86.39
−20	67.74
−15	53.39
−10	42.45
−5	33.89
0	27.20
5	22.05
10	17.96
15	14.68
20	12.09
25	10.00
30	8.313
35	6.941
40	5.828
45	4.912
50	4.161
55	3.537

60	3.021
65	2.589
70	2.229
75	1.924
80	1.669
85	1.451
90	1.366
95	1.108
100	0.9735
105	0.8575
110	0.7579

In this chapter, you will learn how to calibrate a thermistor by fitting its R-T data to a formula known as the Steinhart–Hart Equation. This accomplishment is contingent on the ability to read the calibration data into your fitting program. One method we will explore for inputting data into a VI is through reading a disk file that contains the relevant information.

It would be helpful if, prior to coming to lab, you create and save a spreadsheet-formatted file containing a thermistor's R-T data (either using the provided table or data for your own device). Remember in Chapter 4 you learned that in the spreadsheet-format, tabs separate columns and EOL characters separate rows.

You can use either a spreadsheet data analysis program or a word processor to create the desired file. For example, I used Microsoft Word to construct the ASCII spreadsheet file of the thermistor resistance (in kΩ) versus temperature (in °C) data shown next. When saving the file, it is important to use the **Save As...** command, then select **Text Only** (as opposed to **Normal**) for the **File Type**. This procedure creates an ASCII text file devoid of any Word formatting characters. Take care not to inadvertently add an extra EOL character at the end of the file.

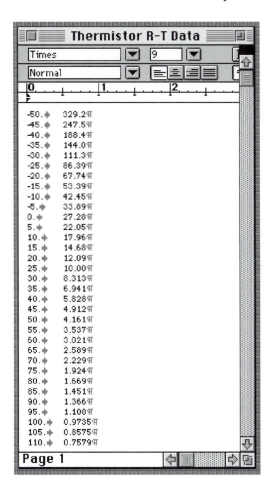

P.225,

Alternately, if you use a spreadsheet-based program to create the file, you will want to save it as a **Text (Tab-Delimited)** file. This option may be listed under a **Save**, **Save As...**, or **Export** menu item, depending on the program you use. Make sure that you save this file in the correct manner to avoid a bunch of program-dependent formatting characters being included in the file. Perhaps the best way to check that you've produced the desired file is to open it using a word processor to see if it appears as above.

Bring your spreadsheet file of R-T data with you to lab. You'll need it during the experiment.

Temperature Measurement Using Thermistors

Among the many electronic temperature sensors available—including platinum resistance thermometers, thermocouples, and semiconductor devices such as the Analog Devices AD590—thermistors are used in applications that require high sensitivity, small size, ruggedness, fast response time, and low cost. When utilized with the proper support circuitry, a thermistor can easily measure temperatures to an accuracy of

0.01°C over a fairly wide range (about 100°C). The only price to be paid for the thermistor's high sensitivity is that its resistance is a nonlinear function of temperature. In the past, the nonlinearity of thermistors was a distinct disadvantage, causing users to employ resistor-based linearizing networks or less-than-accurate resistance-to-temperature conversion methods such as the Beta formula (see following). Fortunately, the widespread availability of microprocessors has enabled the use of accurate (and somewhat complex) temperature-calibration models, greatly simplifying the use of thermistors in accurate temperature measurement.

The active layer of most thermistors is composed of a semiconducting metal-oxide alloy. As is characteristic of a semiconductor, when a thermistor's temperature is elevated, loosely bound (valence) electrons within its active layer are thermally released from their respective binding sites to become conduction electrons. By this mechanism the thermistor's resistance decreases with temperature and results in the utility of this device as a temperature sensor. Since the bound electronic state is of lower energy, an activation energy barrier E separates the conducting state from the insulating state. Thus at a given temperature T (in Kelvin), we expect that the probability of such activation, and therefore the density n of conduction electrons, is proportional to a Boltzmann factor. That is,

$$n = n_0 \exp\left[-\frac{E}{kT}\right] \tag{1}$$

where n_0 is the density of conduction electrons at very high temperatures and k is Boltzmann's constant. It is easily shown that these conduction electrons imbue the material with a resistivity ρ given by

$$\rho = \frac{m}{ne^2\tau} \tag{2}$$

where e and m are the charge and mass of an electron, respectively. The collision time τ is the average time traveled by a conduction electron until it is scattered by either a lattice vibration, impurity, or some other scattering mechanism peculiar to the material.

Thus, if we assume that e, m, and τ are independent of temperature, we expect that the temperature dependence of ρ is entirely due to the Boltzmann factor related to n. Putting (1) into (2), we predict that

$$\rho = \rho_0 \exp\left[\frac{E}{kT}\right] \tag{3}$$

where ρ_0 is the temperature-independent constant given by

$$\rho_0 = \frac{m}{n_0 e^2\tau} \tag{4}$$

The macroscopic resistance R of a sample is determined by

$$R = \rho\frac{L}{A} \tag{5}$$

where L and A are the sample's length and cross-sectional area, respectively.

Because a solid sample's geometric shape is little changed with temperature, we expect that the temperature dependence of R will simply mimic that of ρ. Thus

$$R = R_0 \exp\left[\frac{E}{kT}\right] \tag{6}$$

where R_0 is the (assumed) temperature-independent constant

$$R_0 = \frac{mL}{An_0 e^2 \tau} \tag{7}$$

If all of our logic is correct, we predict that the temperature dependence of a thermistor's resistance should be described by Equation (6). By taking the logarithm of both sides, this relation can be rewritten as the following linear relation

$$\frac{1}{T} = -\frac{k}{E}\ln R_0 + \frac{k}{E}\ln R \tag{8}$$

or

$$\frac{1}{T} = A + B\ln R \tag{9}$$

where A and B are constants defined to be $A \equiv -k\ln R_0/E$ and $B \equiv k/E$. Equation (9) is the so-called *"Beta formula"* (B is sometimes written as β) and it predicts that when a thermistor's temperature-dependent resistance data is plotted as $1/T$ (y-axis) versus $\ln R$ (x-axis), a straight line will result with the y-intercept and slope equal to A and B, respectively.

In spite of this expectation, it is found experimentally that a real thermistor's R-T data exhibits some nonlinearity when plotted as $1/T$ versus $\ln R$. Rather than being constant, the slope B decreases with decreasing temperature. Over a narrow temperature range, say of 10°C, B varies so slightly that the Beta formula can predict the thermistor's behavior with a temperature uncertainty of about 0.01°C. However, for the much larger temperature spans over which a thermistor is useful, the Beta formula inadequately models the data.

Something then must be missing in our theory. In reviewing the above derivation, one might suspect it is our assumption of a temperature-independent collision time. It is easy to imagine that as temperature increases, the conduction electrons move faster and/or lattice vibrations become more pronounced, and consequently shorten the scattering time τ. While the exponential Boltzmann factor activation of conduction electron density will still dominate the resistance's temperature dependence, the more mild temperature-dependent scattering time can possibly account for the observed slight nonlinearity of a $1/T$ versus $\ln R$ plot.

How then can we more accurately model the R-T data? Unfortunately, the theory for electrical conduction in metal oxide thermistors is still incomplete and cannot offer us a more refined relation to replace Equation (9). (In fact, the semiconductor energy band model advocated above is not accepted by all researchers. Some investigators, instead, hold that electric conduction in these materials is due to "hopping" of charge carriers from one ionic site to the next.) In light of this theoretical vacuum, the most recent literature on thermistors accounts for the nonlinearity of the $1/T$ versus $\ln R$ curve by using the standard curve-fitting technique of considering $1/T$ to be a polynomial of $\ln R$. That is, the inverse temperature is written as the following n^{th} order polynomial:

$$\frac{1}{T} = A + B \ln R + C(\ln R)^2 + \ldots + Q(\ln R)^n \qquad (10)$$

From this point of view, Equation (9) is such a polynomial that has been truncated at the first-order term. One then is led to the conclusion that by retaining higher-order terms, the equation will remain valid over a wider range of temperatures.

Starting with this idea, researchers have found that the 100°C temperature span over which a typical thermistor is active can be accurately modeled by a third-order polynomial:

$$\frac{1}{T} = A + B \ln R + C(\ln R)^2 + D(\ln R)^3 \qquad (11)$$

This relation can be used to convert resistance to temperature with a precision equal to that of the original R-T data (typically 0.005 to 0.01°C).

In 1968, the two oceanographers Steinhart and Hart discovered that, when modeling up to 200°C spans contained within the temperature range of –70°C to 135°C (they were most interested in the oceanographic range of –2°C to +30°C), the second-order term in the above relation can be neglected without significant loss of accuracy. Then Equation (11) reduces to

$$\frac{1}{T} = A + B \ln R + D(\ln R)^3 \qquad (12)$$

Equation (12) is called the *"Steinhart-Hart Equation"* and is widely used in thermistor calibration. For a typical thermistor, the constants A, B and D have order of magnitude values of 10^{-3}, 10^{-4} and 10^{-7}, respectively, when resistance is measured in ohms. The temperature must, of course, be on the Kelvin scale.

The Least-Squares Linear Method

The process of scientific inquiry many times proceeds as follows: Researchers perform an experiment to investigate a particular physical phenomenon and to obtain a set of N data points (x_j, y_j). After shutting off their instruments, the investigators apply appropriate curve-fitting techniques to determine the functional relationship $y(x)$ manifest in their data. Once armed with this $y(x)$, the scientific community can then judge the veracity of theoretical models posited to explain the phenomenon by testing each model's capacity for correctly predicting the experimentally observed $y(x)$. In this section, we will explore a curve-fitting method that allows one to extract an accurate $y(x)$ from a given data set (x_j, y_j), a skill that plays a crucial role in the above-described process. We will put this skill to immediate use in calibrating a temperature-sensing thermistor.

Quite often it is convenient to model a functional relationship $y(x)$ as the linear combination of a set of basis functions Z. The general form of this kind of model is

$$y(x) = \sum_{k=0}^{M-1} a_k Z_k(x) \tag{13}$$

where the basis functions Z_k are known functions of x and the a_k are the linear expansion coefficients. For example, if $Z_k(x) = x^k$, then $y(x)$ is a polynomial of order $M - 1$. Alternately, if $Z_k(x) = \cos(k\pi x/L)$, then $y(x)$ is an even function described by a Fourier series. Note that the functions $Z_k(x)$ can be nonlinear functions of x. However, the coefficients a_k appear in a linear fashion in Equation (13).

Once the decision is made to model $y(x)$ as a linear combination of a set of basis functions, one needs some method to determine the appropriate choice of the a_k in Equation (13), so that a given data set is accurately described. The theory of data analysis provides just such a method. Because all data-taking processes are subject to random errors, the repeated acquisition of a particular experimental quantity produces an array of values that is distributed as a Gaussian ("bell curve") distribution. By carefully considering the ramification of this fact, it can be shown that Equation (13) will best describe a given N data points (x_j, y_j) when the a_k are chosen such that they minimize a "merit-of-fit" function defined as

$$\chi^2 = \sum_{j=0}^{N-1} \left[\frac{y_j - \sum_{k=0}^{M-1} a_k Z_k(x_j)}{\sigma_j} \right]^2 \tag{14}$$

where σ_j is the measurement uncertainty (standard deviation) of the jth data point. Various algorithms have been proposed to identify the a_k which minimizes χ^2 for a given data set, all of which are classified under the rubric of *"least-squares linear "* methods. In most applications, the Singular Value Decomposition (SVD) method works best. Other methods include the Givens, Householder, LU (Lower Triangular-Upper Triangular) Decomposition and Cholesky algorithms. The interested reader may refer to a data analysis text such as *Numerical Recipes* (Cambridge University Press, 1986) for a description of the details of these methods.

LabVIEW provides built-in VIs that will implement a least-squares linear fit of your data to the functional form of Equation (13). In the format we will discuss, you supply the VI with your data and the desired basis functions Z_k. The VI, through the magic of least-squares methodology, determines and outputs the coefficients a_k that best fit your experimental results.

How then do you make known your choice of basis functions to the VI? This communication, unfortunately, takes some work on your part. Remember, we have two givens:

N data points (x_j, y_j) $j = 0, 1, 2, ..., N - 1$

M basis functions $Z_k(x)$ $k = 0, 1, 2, ..., M - 1$

where $Z_k(x)$ are known analytic functions. Starting with these two givens, LabVIEW asks you to construct what is known as a *"design matrix"* H, which is defined to be the following N × M matrix:

$$H = \begin{bmatrix} Z_0(x_0) & Z_1(x_0) & Z_2(x_0) & \cdots & Z_{M-1}(x_0) \\ Z_0(x_1) & Z_1(x_1) & Z_2(x_1) & \cdots & Z_{M-1}(x_1) \\ Z_0(x_2) & Z_1(x_2) & Z_2(x_2) & \cdots & Z_{M-1}(x_2) \\ \vdots & & & & \vdots \\ Z_0(x_{N-1}) & Z_1(x_{N-1}) & Z_2(x_{N-1}) & \cdots & Z_{M-1}(x_{N-1}) \end{bmatrix} \tag{15}$$

The H matrix is the actual input to the least-squares linear fitting VI and thus it is the medium through which the VI is alerted to your choice of basis functions.

So, for example, let's say you were trying to fit a thermistor's R-T data to the Steinhart–Hart Equation. Then in the language of Equation (13), $y = 1/T$ and $Z_0 = 1$, $Z_1 = \ln R$ and $Z_2 = (\ln R)^3$. Therefore, given N data points (R_i, T_i), the H matrix will be

$$H = \begin{bmatrix} 1 & \ln R_0 & (\ln R_0)^3 \\ 1 & \ln R_1 & (\ln R_1)^3 \\ 1 & \ln R_2 & (\ln R_2)^3 \\ \vdots & & \vdots \\ 1 & \ln R_{N-1} & (\ln R_{N-1})^3 \end{bmatrix} \tag{16}$$

INPUTTING DATA TO A VI USING A FRONT-PANEL CONTROL

In the following pages, you will write a program that fits a thermistor's R-T data to the Steinhart–Hart Equation. In order to write this VI, you must first understand how to input the given array of R-T data into a LabVIEW program.

The most straightforward method for inputting an array of values is from a front-panel Array Control. Let's write a VI that works this way. Construct the front panel shown here. Place two Array Controls (select an **Array** shell first, then place a **Digital Control** inside) and label one **Resistance (kilohms)**, the other **Temperature (deg C)**. Using its pop-up menu, increase the **Resistance (kilohms)** control's **Digits of Precision** to *4* (or more, if needed). Finally, add an **XY Graph** labeled **R vs. T** so that the thermistor data may be viewed graphically. Save the VI as **R-T Plot-Array Controls** in **YourName.llb**.

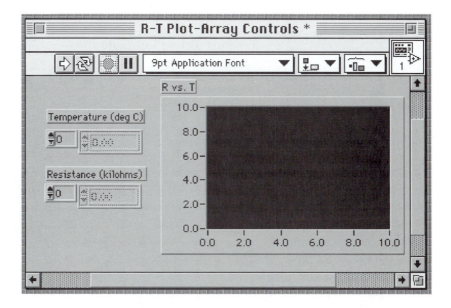

Switch to the block diagram. Write the following code that receives the input arrays and outputs them in the appropriate manner for plotting on the **XY Graph**.

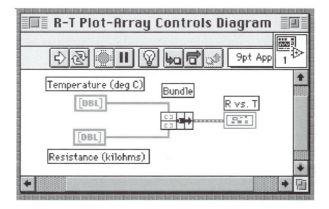

Return to the front panel. Here's the hard part. Manually type the N temperature values from your thermistor calibration data table into the **Temperature (deg C)** control. Then type the N resistance values from your data table into the **Resistance (kilohms)** control. Remember that the small left-hand box portion of the Array Control (called the *"index display"*) indicates the index of the element being displayed, and the larger right-hand box (the *"element display"*) is the actual numerical value of that element. You increment or decrement the index display using the Operating Tool. Also use this tool to highlight the interior of the element display, then type in the numerical value that you desire.

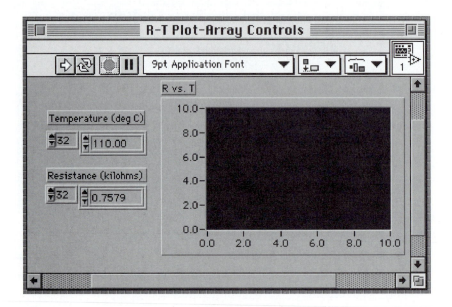

Now we want to store the information in these two arrays permanently. That is, we wish for the arrays to be initialized with these values each time the VI is opened. To accomplish this feat, select **Make Current Values Default** in the **Operate** menu.

Run the VI and you should see a graph of your thermistor's resistance as a function of temperature. Note the exponential-looking decay of resistance with increasing temperature. This is a signature of the thermally activated process taking place within the material of which the thermistor is composed.

Try closing, then opening the VI. Repeat this process a few times. Hopefully, you'll find the two front-panel arrays are initialized with the thermistor data each time the program is loaded.

Entering data manually into the arrays was hard work. Now that that task is complete, you may worry that someone, or (in a moment of ineptitude) you, might accidentally modify the information in these arrays. "If only I could hide these arrays from prying hands," you say to yourself, "then my data would be safe." Well, the LabVIEW creators have anticipated your wishes.

Switch to the block diagram and pop up on the **Resistance (kilohms)**'s terminal. Select **Hide Control** from the menu. Repeat this process at the **Temperature (deg C)** terminal.

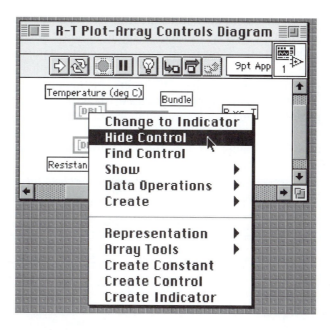

Return to the front panel. You'll find that the array controls are now hidden and out of harm's way! Try running the VI to assure yourself that the VI still can access the data.

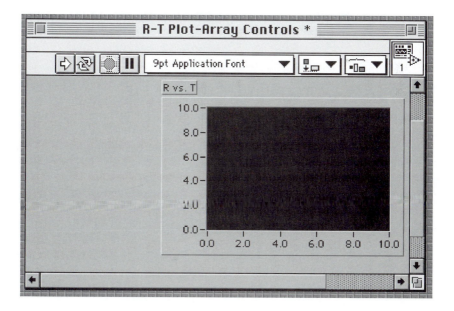

Save your final version of this VI as you close it.

You can also input data into a program by placing an array directly on the block diagram using the **Array Constant** icon found in **Functions>>Array**. The **Array Constant**, just like the front-panel **Array** shell, comes equipped with an index display as well as an element display receptacle that you stuff with a constant of desired data-type (numeric, Boolean, string, or cluster). In this approach, you first position an **Array Constant** on your block diagram, then place a **Numeric Constant** from **Functions>>Numeric** (with Representation **DBL**) within its "element receptor." You then program the **Array Constant** with the relevant sequence of data values. The resulting block diagram for **R-T Plot–Array Controls** would appear as below.

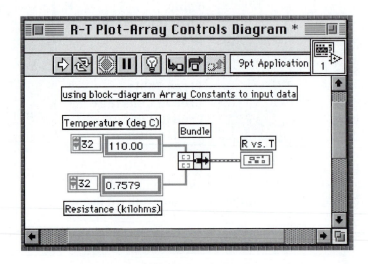

INPUTTING DATA TO A VI BY READING FROM A DISK FILE

An alternate method for inputting data into a VI is by reading it in from a previously created disk file. Let's write a LabVIEW program that can receive and plot the thermistor spreadsheet data that you generated prior to coming to lab. If you haven't created this file yet, now is the time to do it.

Construct this front panel, which simply contains an **XY Graph**. Save this VI in **YourName.llb** under the name **R-T Plot–Spreadsheet**.

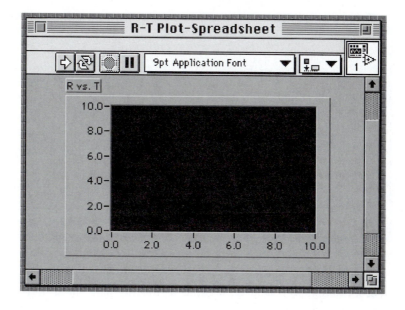

Switch to the block diagram. An ASCII spreadsheet file can be read into your VI using the **Read From Spreadsheet File.vi** icon found in **Functions>>File I/O**. The *Simple Diagram Help Window* for this icon is shown in the following illustration. As with all of the more sophisticated LabVIEW VIs, **Read From Spreadsheet File.vi** can be used for a lot of specialized purposes and therefore has many input and output connections. All of these available connections are described on the *Detailed Diagram Help Window*, which can be accessed by clicking on the button in Help Window's bottom left-hand corner. However, for most applications (this time, ours included), you need only connect wires to the input and outputs displayed in the Simple Diagram Help. The inputs and outputs that appear in boldface (**Read From Spreadsheet File.vi**

has none of these) are those that must be specified. All other terminals denote features available for your use, if desired. Default values for inputs are shown in parentheses.

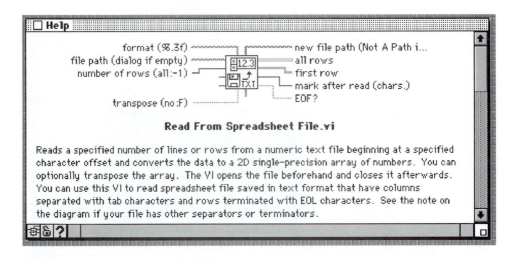

Place this icon on your block diagram and wire a **Path Constant** containing your spreadsheet-formatted R-T data file's name to the **file path** input. If you leave this input unwired, a dialog box will appear when the program is run and prompt you for the desired file.

Also, increase the precision of numerical values from the insufficient default value of *3* decimal places to *4* (or more, if needed) by wiring a **String Constant** containing %.*4f* to the **format** input. Finally, since we want to read the entire file, which is the default value for the **number of rows** input, leave this input unwired.

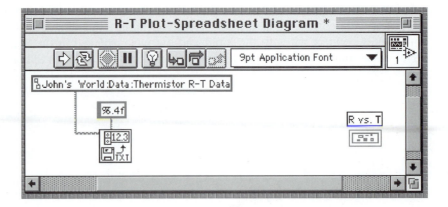

Slicing Up a Multi-Dimensional Array

Read From Spreadsheet File.vi will present your data at its **all rows** output in the form of a 2D array. This array's two columns then will have to be separated ("sliced off") into two 1D arrays to facilitate an XY plot. Separating out the columns is done using the **Index Array** icon. Place this icon on your diagram and use the Positioning Tool to stretch it so that it includes two index inputs.

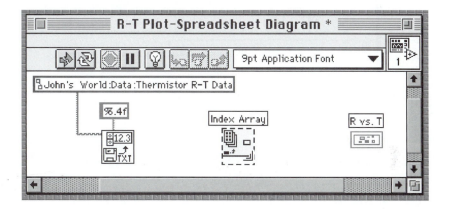

The top and bottom index inputs, which appear as small black boxes, are the row and column indices, respectively. First, let's retrieve the temperature values from the 2D array. We know that the temperature values are all contained in the index-zero (first) column. So wire Read From Spreadsheet File.vi's **all row** output to Index Array's **n-dimension array** input, then wire a **Numeric Constant** containing a zero (*0*) to the column (bottom) index input.

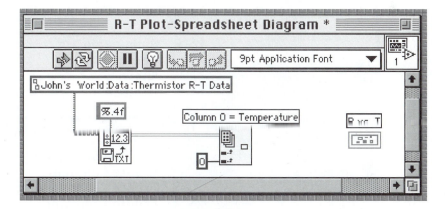

Now if we, for example, wanted to pick out the eighth temperature value in this column, which has the index of seven, we would wire a **Numeric Constant** containing 7 to the top index input. **Index Array** would then output this single array element. However, instead of a single temperature value, we want the entire column of temperature values. To instruct **Index Array** to "slice off" the entire column, we will disable the row index, so that all rows in the selected column are output in the form of a 1D array. To disable the row indexing, place the Positioning Tool over the top (row) index input, as shown.

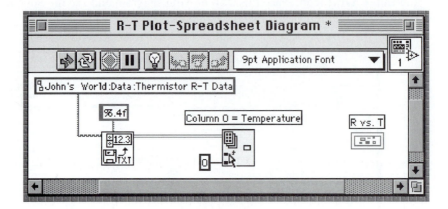

Then pop up its menu and select the **Disable Indexing** option.

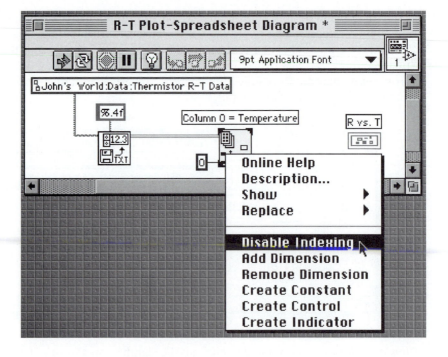

The row index input will then appear as a hollowed-out box, indicating that indexing at this input has been toggled off.

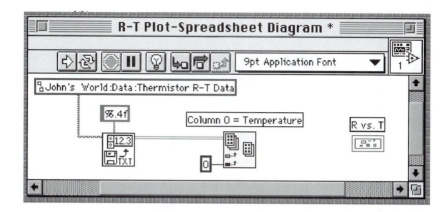

Repeat the above procedure to slice off the resistance data, which is in the index-one (second) column.

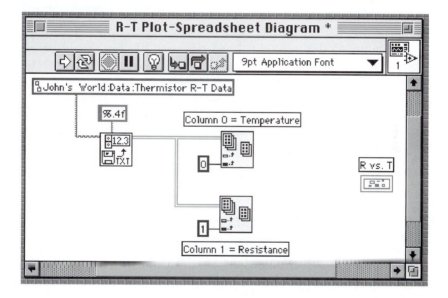

The **Index Array** icons below will each output a 1D array. Bundle these two 1D arrays together to form an XY cluster and wire this cluster to the XY Graph's terminal.

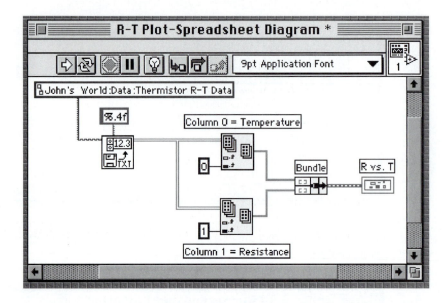

Return to the front panel and run your VI. You should be presented with a beautiful graph of your thermistor's resistance (y-axis) versus temperature (x-axis). If you encounter an error when your VI tries to open the data file, carefully check that the file name within the block diagram's **Path Constant** is perfectly correct. Else, check to make sure your data file is correctly constructed—such an error may be easily detected by running your program with strategically placed Probes and/or by using Highlight Execution. An extraneous EOL character at the end of the data file is a common problem.

Save your work as you close this VI.

CURVE FITTING USING THE LEAST-SQUARES LINEAR METHOD

Now that you know how to input your thermistor data into a program, let's write a VI that fits these data to the Steinhart–Hart Equation [Equation (12)]. The fitting will be done by the **General LS Linear Fit.vi** icon, which is found in **Functions>>Analysis>> Curve Fitting**.

The Help Window for **General LS Linear Fit.vi** is shown here. This VI finds the best fit of experimental data (x_j, y_j) to the linear combination of a set of basis functions $Z_k(x)$. That is, the data input to this icon is fit to Equation (13).

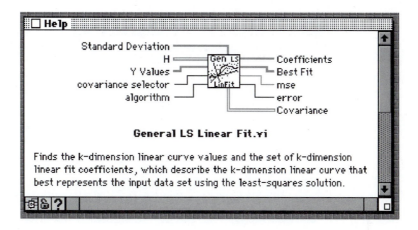

Assuming that the uncertainty σ_j is the same for each of the measured quantities, you do not have to input anything to the **Standard Deviation** array input. The **algorithm** input selects which technique will be used to fit the data according to the following code: 0 = SVD, 1 = Givens, 2 = Givens2, 3 = Householder, 4 = LU decomposition, 5 = Cholesky. If this input is left unwired, the icon will use the SVD method by default. The SVD method is the best generic choice because it always produces solutions (the others fail sometimes), although it may have a speed and memory disadvantage in some cases. By inputting the H matrix, which communicates your choice of basis functions to the fitting algorithm, and the array of experimental y-values, the **General LS Linear Fit.vi** icon outputs the optimized expansion coefficients a_k and the array of *"Best Fit" y-values* where

$$y_j^{\text{"Best Fit"}} \equiv y(x_j) = \sum_{k=0}^{M-1} a_k Z_k(x_j) \tag{17}$$

The *mean square error (mse)* is also output, which is defined as

$$mse \equiv \frac{1}{N} \sum_{j=0}^{N-1} (y_j - y_j^{\text{Best Fit}})^2 \tag{18}$$

The square root of *mse* can be taken as a measure of the error in your fitted curve.

Let's first write two VIs—**Data Input** and **H Matrix**—which will be used as subVIs in your final program. **Data Input** will read in the experimental data and **H Matrix** will construct the design matrix.

Build the **Data Input** front panel shown below using four Array Indicators. Also use the Icon Editor to design an icon, then assign the connector's four terminals consistent with the Help Window shown. Save this work in **YourName.llb**.

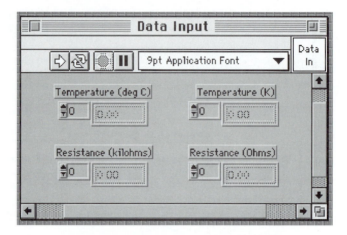

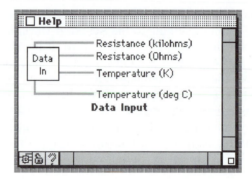

Now write the following block diagram. This VI reads the thermistor's R-T data and outputs the temperature and resistance in the two differing units that you will require in your main program. Run the VI to verify that it works.

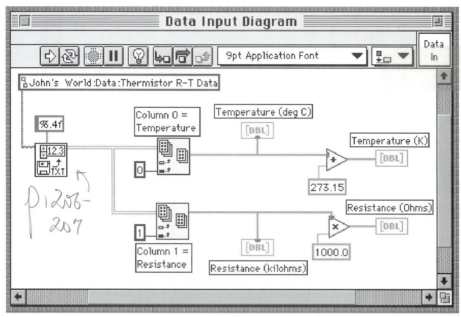

Now write **Build H**, a VI that accepts the thermistor's resistance data as input and processes it to output the design matrix needed for fitting the R-T data to the Steinhart–Hart Equation. The appropriate design matrix is given by Equation (16).

Construct the following front panel with a 1D Array Control **Resistance (Ohms)** and a 2D Array Indicator **H Matrix**. See the following discussion to find out how to create a 2D Array Indicator. Design the icon and assign the two connector terminals as shown in the Help Window.

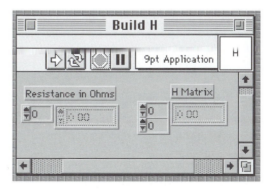

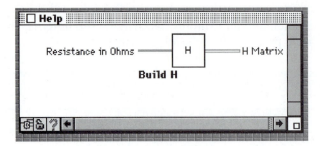

Constructing a 2D Array Indicator

To make a 2D Array Indicator, first place a 1D Array Indicator on the panel.

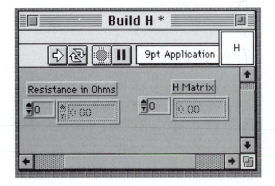

Then pop up on the index display and select **Add Dimension**. The Array Indicator will now be associated with a 2D array and so two index-display boxes will appear. The top and bottom index displays select the row and column indices, respectively, for the array element whose numerical value is shown in the element display.

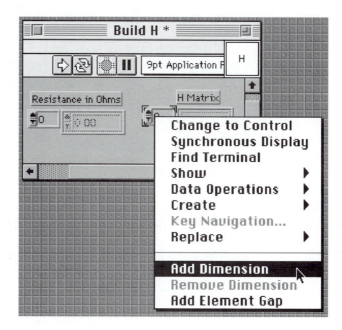

Alternately, you can create the same 2D object by the following easier method: Place the Positioning Tool in the corner of the 1D Array Indicator's index display so that it morphs into a *grid cursor* ▦.

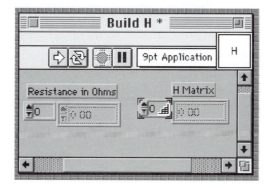

Then drag the grid cursor down until an additional index display appears.

When you release the mouse button, a row and column index display will be attached to the Array Indicator, indicative of its transformation to a two-dimensional object.

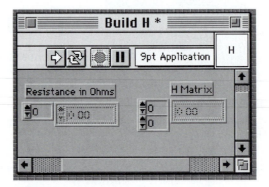

Code the block diagram as shown next. The experimental resistance values R_j are input one by one to the For Loop using the loop's (default) auto-indexing feature. Given R_j during the jth loop iteration, the Formula Node evaluates the Steinhart–Hart Equation's three basis functions, then assembles them in a three-column "row-like"

array. At its output, the loop's auto-indexing feature builds an N-row × 3-column 2D array, whose j^{th} row is the row-like array assembled during the jth loop iteration. Save this VI in **YourName.llb**.

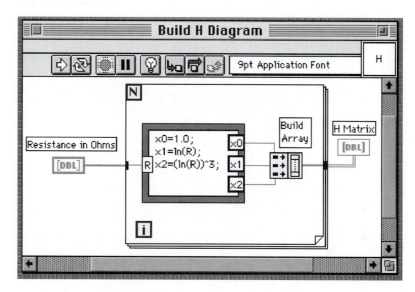

p.231

Finally, you are ready to write your main program, which will fit the thermistor's experimental data to the Steinhart–Hart Equation. Build the front panel given below. To compare the resulting fit with the original data, include an **XY Graph** with two differentiated plots (for example, select different colors or point styles in the Legend), the first called **Data** and the second called **Fit**. Also, display the fitted values of a_k in an Array Indicator labeled **Best Coefficients** and the "fitting error" in a **Digital Indicator** called **Square Root of mse**. Pop up on these indicators and greatly increase their digits of precision, say, to *12*. Save this panel as **R-T Fit and Plot** in **YourName.llb**.

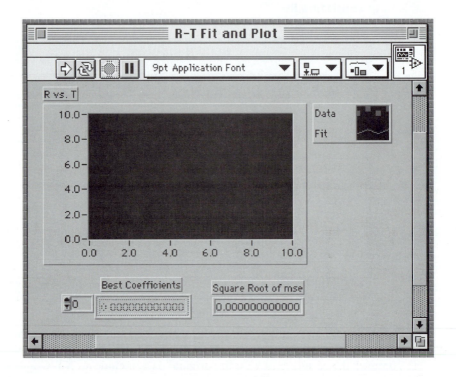

Switch to the block diagram. Write the following program that implements **Data Input** to read in the thermistor's experimental R-T data. The R_j values (in ohms) are then fed to **Build H**, which constructs the design matrix, and supplies this 2D array to the **H** input of the **General LS Linear Fit.vi** icon. Also, the 1D array of Kelvin temperatures is input to the **Reciprocal** icon (found in **Functions>>Numeric**), which acts on each array element separately to produce an array of inverse Kelvin temperatures for the General LS Linear Fit.vi's **Y Values** input. Following the fitting algorithm, the **Best Coefficients** and **Square Root of mse** are output.

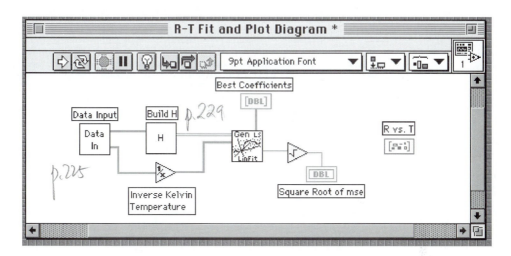

So that you're aware of its existence, I'll mention that you could also form the array of inverse temperatures by using **Power of X**, found in **Functions>>Numeric>>Logarithmic**, as shown in the following block diagram.

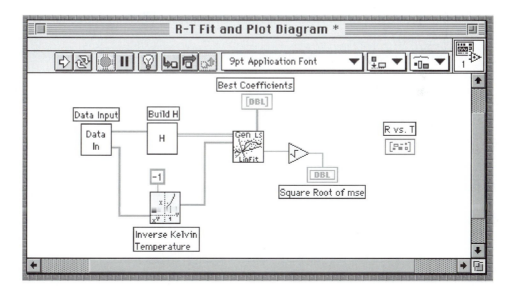

Finally, add the following code that plots the original data and the fitted results for comparison. General LS Linear Fit.vi's **Best Fit** output contains the array of $y_j^{\text{"Best Fit"}}$ values. Since y = 1/T, where T is in Kelvin, this array must be converted to Celsius temperatures prior to the comparison plot.

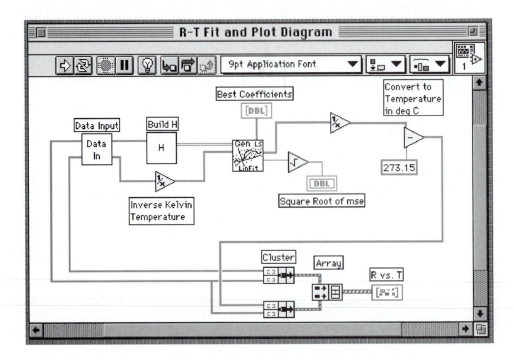

Return to the front panel and run your VI. Upon completion, the **Best Coefficients** array indicator will display the Steinhart-Hart's A, B, and D constants at its *0*, *1*, and *2* indices, respectively.

Creating a Multi-Element Array Indicator

For convenient reading, you can make all three array elements visible at once, if you like, through use of a grid cursor. Here's the procedure: First, place the Positioning Tool in the corner of the **Best Coefficients** indicator on the front panel so that it morphs into a grid cursor.

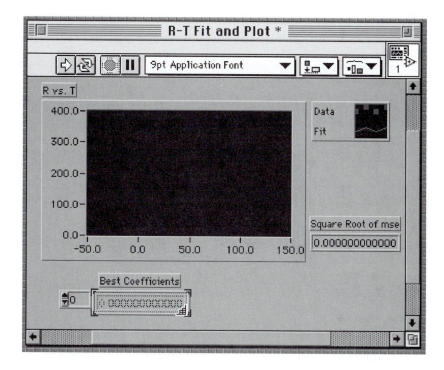

Then drag the grid cursor to the right, until it has created two additional indicators.

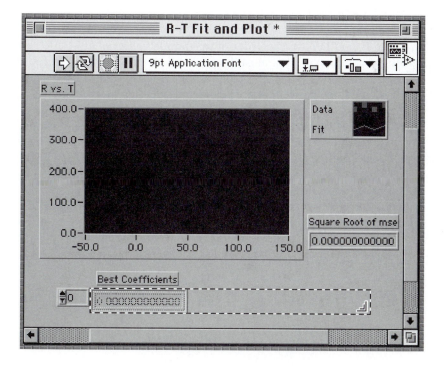

Release the mouse button and indicators for array elements *0, 1*, and *2* will now be visible. The index display provides the index for the element displayed in the leftmost indicator.

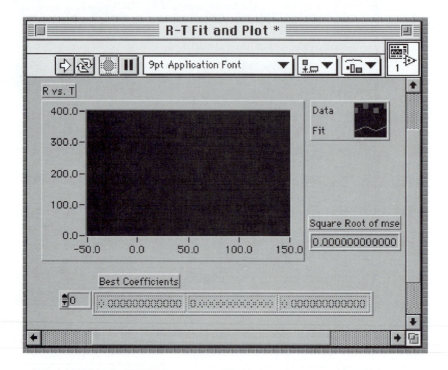

Record the resulting values of A, B, and D for your thermistor for future use. Do your A, B, and D have the order of magnitude values of 10^{-3}, 10^{-4}, and 10^{-7}, respectively, which are typical of most thermistors? Given the theoretical considerations discussed at the beginning of this chapter, can you use your value of B to determine the activation energy barrier E for electronic conduction in your thermistor (Boltzmann's constant $k = 8.6 \times 10^{-5}$ eV/K)? Does the **XY Graph** indicate that the fitted Steinhart–Hart Equation accurately describes the temperature-dependent resistance of your thermistor? Finally, record the value of the **Square Root of mse**. This value can be interpreted as the precision (in units of inverse Kelvin degrees) of your fit.

When curve fitting, one expects that a better fit can be achieved by increasing the number of basis functions used in the fitting algorithm. Thus, you might predict that a third-order polynomial [Equation (11)] will fit your thermistor data better than the Steinhart–Hart Equation [Equation (12)], which deletes the second-order term. Let's see if this conjecture is true. To fit your data to Equation (11), open **Build H** and modify it to include the four necessary basis functions, as shown next.

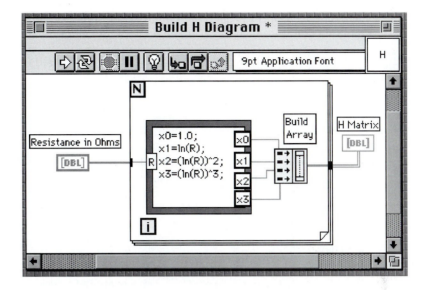

With this modification in place, run **R-T Fit and Plot**. Record the new value of **Square Root of mse**. Has the accuracy of your fit dramatically improved or is your result consistent with Steinhart–Hart's claim that the second-order term can be deleted from the fitting equation without significant loss of accuracy? Note that by adding the $(\ln R)^2$ term, the new coefficients for the 1, $\ln R$, and $(\ln R)^3$ terms become different from those derived in the Steinhart–Hart case. This phenomenon results from the fact that the polynomials x^k are not an *orthogonal* basis set. Take a linear algebra (or quantum mechanics) course for more details.

Finally, try fitting your data to the Beta formula [Equation (9)]. In this first-order polynomial fit, only two basis functions are required. Modify **Build H** appropriately, as shown here.

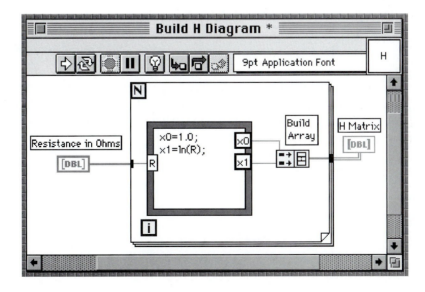

Now run **R-T Fit and Plot** and note the resulting value of **Square Root of mse**. You should find that it is much larger than the previous values found, indicative of the Beta formula's deficiency. The deviation of the fitted function from the data may be apparent on the **XY Graph**, especially at the lower temperatures.

Built-In Curve Fitting VIs

For future reference, LabVIEW provides built-in VIs that perform a least-squares fit of data to commonly used equations including a straight line, an exponential curve and an mth order polynomial. You will find **Linear Fit**, **Exponential Fit**, and **General Polynomial Fit** in **Functions>>Analysis>>Curve Fitting**.

As an illustrative example of how these curve-fitting VIs function, consider the Help Window for **General Polynomial Fit.vi**:

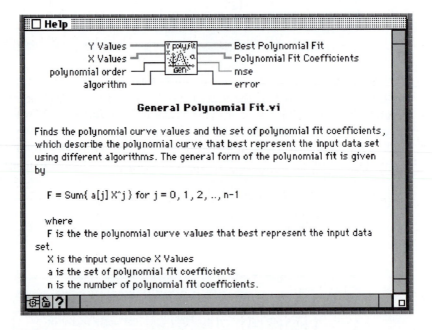

Here, we find that one simply supplies some (x,y) data in the form of x and y arrays and also indicates the order of polynomial to which the data is to be fit. The icon then conveniently outputs the desired polynomial coefficients. In the next illustration, the block diagram for **General Polynomial Fit.vi** is shown and, based on your work in this chapter, you should have no problem understanding how it works. The parameter choice at the **algorithm** input is made using the terminal of an **Enumerated Type** control ⬛. We will learn to use the **Enumerated Type**, a handy control for selecting from a menu of options, in subsequent chapters.

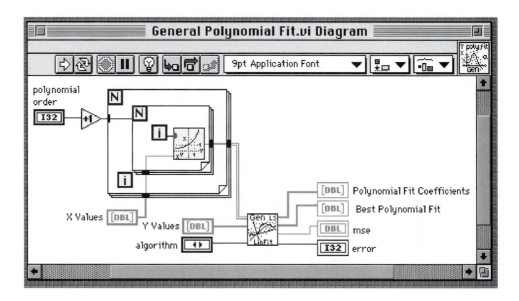

The nested For Loops enclosing **Power of X** create an H matrix appropriate for a polynomial fitting function (that is, using the basis functions x^k), with the outer loop creating rows and the inner loop building the columns for each particular row. Once constructed, the 2D H matrix is passed to the **General LS Linear Fit.vi** icon where the fitting algorithm is implemented. Other curve-fitting VIs such as **Linear Fit** and **Exponential Fit** differ simply in the choice of basis functions used in the construction of the H matrix.

CURVE FITTING USING THE LEAST-SQUARES NONLINEAR METHOD

Finally, you might find it interesting to investigate the more general **Nonlinear Lev-Mar Fit.vi** (available within **Functions>>Analysis>>Curve Fitting**), which implements a least-squares fit via the Levenberg–Marquardt method. This method can fit a data set to a given functional form y = f(x), even if f includes the fitting coefficients in a nonlinear fashion. The Help Window for **Nonlinear Lev-Mar Fit.vi** is given here.

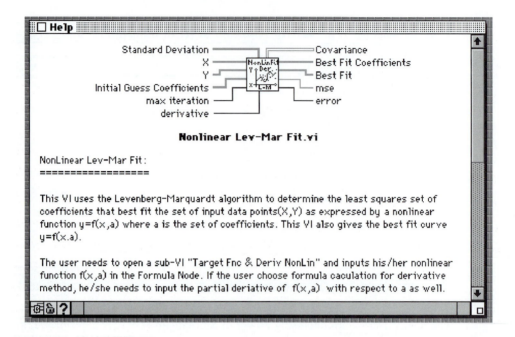

Try fitting your thermistor data to the Steinhart–Hart Equation (which is a linear relationship between the quantities of inverse temperature and powers of the natural logarithm of resistance) using a block diagram based on the **Nonlinear Lev-Mar Fit.vi** icon. Open **R-T Fit and Plot**, then using **Save As…** create a new VI called **R-T Fit and**

Plot Using Lev-Mar in **YourName.llb**. Add two new front-panel controls, **Initial Guess Coefficients** (which is an Array Control) and **Derivative (I32)**.

Then modify the block diagram so that, after the thermistor data is read from a spreadsheet file, the natural logarithm of resistance and inverse temperature are input to the

x and **y** inputs of **Nonlinear Lev-Mar Fit.vi**, respectively. Then wire the new front-panel control terminals to their appropriate inputs.

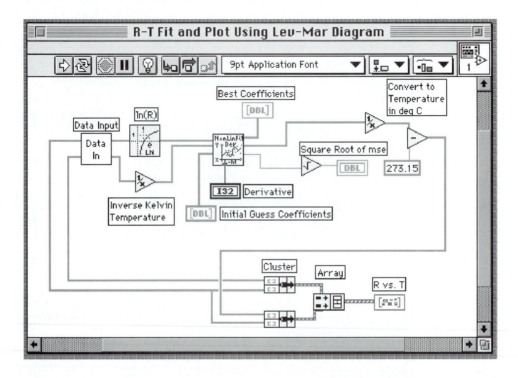

Within a subVI of **Nonlinear Lev-Mar Fit.vi** called **Target Fnc & Deriv NonLin.vi**, the user inputs an analytic expression for the (not-necessarily linear) "target" fitting function, along with its derivative with respect to each of the fitting coefficients. You have two choices of how to do the differentiation, selectable by the icon's **derivative** input. If **derivative** equals *0*, you provide only an analytic expression for the target function and the VI numerically differentiates it. If **derivative** equals *1*, you provide analytic expressions for both the target function and its derivatives with respect to each fitting coefficient. When the target function is easy to differentiate analytically, the latter choice is preferable because it yields faster and more accurate results.

Use **Unopened SubVIs** in the **Project** menu to open **Target Fnc & Deriv NonLin.vi**, then edit its block diagram as shown below. With $x = ln(R)$ and $y = 1/T$, the target function for this fit is $f(x) = a_0 + a_1*x + a_2*x^3$ (Steinhart–Hart Equation). The three derivatives requested are $d_0 \equiv \dfrac{\partial f}{\partial a_0} = 1$, $d_1 \equiv \dfrac{\partial f}{\partial a_1} = x$, $d_2 \equiv \dfrac{\partial f}{\partial a_2} = x^3$.

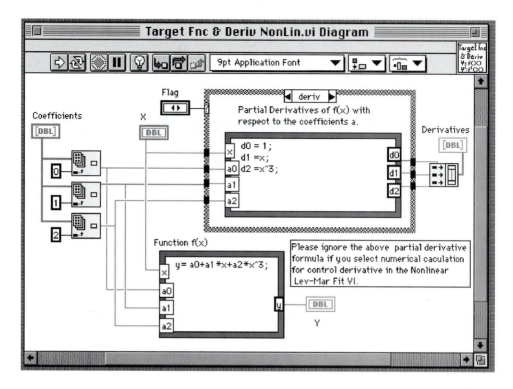

Return to the front panel of **R-T Fit and Plot Using Lev-Mar**. Input "initial guess" choices for the coefficients a_0, a_1, and a_2 using the Array Control **Initial Guess Coefficients**. Selecting **Derivative** equal to *0* or *1* will determine whether **Nonlinear Lev-Mar Fit.vi** performs its required differentiation processes numerically or using the supplied analytic expressions, respectively. The success of the Levenberg–Marquardt method sometimes depends on how close your initial guess coefficients are to the actual solution. Therefore, run your program in an iterative way (that is, using the **Best Coefficients** from one run as the **Initial Guess Coefficients** for the next run) until you have converged on an accurate set of "best" values for the coefficients a_0, a_1, and a_2. Are your values (to a high degree of precision) identical to those obtained from **General LS Linear Fit.vi**?

Here, for the sake of gaining practice in the use of this VI, we applied **Nonlinear Lev-Mar Fit.vi** in a circumstance where data obeys a linear relation. The real utility of the Levenberg Marquardt method, however, is that it will successfully curve fit data that conform to a nonlinear model. One such (frequently occurring) model is the Gaussian distribution where two experimentally measurable quantities x and y are related by a function in which the coefficients A, $\bar{x}$, and σ appear in a nonlinear manner:

$$y(x) = A\exp\left[-\frac{(x - \bar{x})^2}{2\sigma^2} \right]$$

Using **Nonlinear Lev-Mar Fit.vi** to fit (x,y) data from such a system, the experimenter could extract values for the constants A, $\bar{x}$, and σ which correspond to the maximum value of y, the mean value of x, and the standard deviation of the distribution, respectively.

Built-In Analysis VIs—Fast Fourier Transform

The Fourier Transform

Imagine you are investigating a physical phenomenon by measuring the analog quantity x that varies continuously with time t, that is, $x = x(t)$. Based on our discussion at the beginning of the previous chapter, we expect that $x(t)$ can be modeled as the linear combination of a set of M basis functions Z such that

$$x(t) = \sum_{k=0}^{M-1} a_k Z_k(t) \tag{1}$$

The Fourier transform is an example of just such a model. The seminal work of nineteenth century mathematician Baron Jean Baptiste Joseph Fourier, with subsequent elaboration by others, showed that the infinite set of complex exponentials $\exp(i2\pi ft)$, where the f are all the frequencies between $-\infty$ and $+\infty$, is a complete basis set with which one can form linear combinations to model any arbitrary function. Since an infinite number of exponentials with infinitesimally close frequency spacing is required to model an arbitrary function, the summation in Equation (1) becomes an integral, yielding

$$x(t) = \int_{-\infty}^{+\infty} X(f) e^{i2\pi ft} df \tag{2}$$

where the "*Fourier component*" $X(f)$ determines the amplitude (I'll be much more precise in my use of the word *amplitude* shortly) of the exponential with frequency f. Generally, $X(f)$ is a complex-valued function, carrying both *magnitude* and *phase* information, a point we will carefully consider in this chapter.

Conversely, the Fourier components $X(f)$ may be obtained by the following equation

$$X(f) = \int_{-\infty}^{+\infty} x(t) e^{-i2\pi ft} dt \tag{3}$$

Equations (3) and (2) are called the *Fourier transform* and the *inverse Fourier transform,* respectively.

Discrete Sampling and the Nyquist Frequency

In a real experiment, one cannot sample the quantity x continuously, but must instead settle for discretely sampling it N times. If the samples are taken at equally spaced times, where the time difference between adjacent samples is Δt, then the sampling frequency f_s is given by

$$f_s = \frac{1}{\Delta t} \tag{4}$$

and the N consecutive sampling times are

$$t_j = j\Delta t \qquad j = 0, 1, 2, ..., N-1 \tag{5}$$

Given such a sampling scheme, there is a maximum frequency sine wave that one can expect to detect. This limitation is simply due to the fact that to observe a sine signal one must at the very least, for example, first sample its positive peak, then its negative trough at the next sample, followed by its positive peak at the third sample, and so on. Thus, the minimal sampling of a sine wave is two sample points per cycle. In an experiment then, where your instruments are acquiring data every Δt seconds (at a sampling rate $f_s = 1/\Delta t$), the maximum frequency sinusoidal signal you can detect has a period $T = 2\Delta t$. This detection limit is called the *Nyquist critical frequency* f_c and is given by

$$f_c = \frac{1}{2\Delta t} = \frac{f_s}{2} \tag{6}$$

Additionally, by sampling a waveform x(t) at N equally spaced times t_j, one can only expect to determine the Fourier components X(f) of N equally spaced frequencies f_k in the Fourier transform. This statement follows from the tacit assumption that x(t) is periodic such that $x(t) = x(t + N\Delta t)$, a criterion that forces the waveform x(t) to be composed only of component frequencies that fit an integer number of cycles into the sequence of N samples. The constant (zero-frequency) function obviously meets this criterion. The next lowest acceptable frequency has a period of $N\Delta t$ and thus a frequency of $f_1 = 1/N\Delta t\,(= f_s/N)$. Within the acceptable range of $-f_c$ to $+f_c$, the rest of the frequencies f_k are the integer multiples of f_1. Thus, defining the spacing between adjacent frequencies to be $\Delta f = f_s/N$, the N equally spaced frequencies f_k determined by discretely sampling data are

$$f_k = k\left(\frac{f_s}{N}\right) = k\,\Delta f \qquad k = -\frac{N}{2} + 1, ..., 0, ..., +\frac{N}{2} \tag{7}$$

Although Equation (7) appears deficient in its neglect of the frequency $-f_c$ (given by $k = -N/2$), the two extreme frequencies $\pm f_c = f_s/2$ actually describe the same basis function due to the periodic nature of x(t). Thus the range of k-values in the above relation rightfully runs from $-N/2+1$ to $+N/2$.

The Discrete Fourier Transform

With the above-described constraints due to discrete sampling, we then approximate the Fourier transform integral [Equation (3)] by the following sum:

$$X(f_k) = \int_{-\infty}^{+\infty} x(t)\, e^{-i2\pi f_k t}\, dt \approx \sum_{j=0}^{N-1} x(t_j)\, e^{-i2\pi f_k t_j}\, \Delta t \tag{8}$$

Note, using Equations (4), (5) and (7),

$$f_k t_j = (k\,\Delta f)(j\,\Delta t) = k\, j\, \frac{f_s}{N}\Delta t = \frac{k\, j}{N} \tag{9}$$

Putting (9) into (8), and realizing that Δt is a constant, we find

$$X(f_k) \approx \Delta t \sum_{j=0}^{N-1} x(t_j)\, e^{-i2\pi jk/N} \equiv \Delta t\, X_k \tag{10}$$

where we have defined the *discrete Fourier transform* to be

$$\boxed{\; X_k \equiv \sum_{j=0}^{N-1} x(t_j)\, e^{-i2\pi jk/N} \qquad k = -\frac{N}{2} + 1,\, \ldots,\, 0,\, \ldots,\, +\frac{N}{2} \;} \tag{11}$$

Finally, we approximate the inverse Fourier transform integral [Equation (2)] by the following sum:

$$x(t_j) = \int_{-\infty}^{+\infty} X(f)\, e^{i2\pi f t_j}\, df \approx \sum_{k=0}^{N-1} X(f_k)\, e^{i2\pi f_k t_j}\, \Delta f \tag{12}$$

Putting (10) into (12),

$$x_j \approx \sum_{k=0}^{N-1} X_k\, \Delta t\, e^{i2\pi f_k t_j}\, \Delta f \tag{13}$$

Then using Equation (9) and also noting that $(\Delta t)(\Delta f) = (1/f_s)(f_s/N) = 1/N$, Equation (13) becomes the *discrete inverse Fourier transform*

$$\boxed{\; x_j \approx \sum_{k=0}^{N-1} \frac{X_k}{N}\, e^{i2\pi f_k t_j} = \sum_{k=0}^{N-1} \frac{X_k}{N}\, e^{i2\pi kj/N} \equiv \sum_{k=0}^{N-1} A_k\, e^{i2\pi kj/N} \;} \tag{14}$$

Equations (11) and (14) provide us with a method of performing spectral analysis on an experimental signal x. Given that x has been sampled discretely at N evenly spaced times t_j, we first calculate the N values of X_k using the definition of the discrete Fourier transform [Equation (11)]. Then, (in the spirit of Fourier) viewing the signal x as really the composite of oscillations at N different frequencies f_k, Equation (14) tells us that the amplitude of oscillation A_k at each f_k is given by

$$A_k = \frac{X_k}{N} \tag{15}$$

Since A_k is derived from the complex-valued X_k, in general, A_k is a complex-valued function. Let's then name A_k the *complex-amplitude*. The plot of complex-amplitude A_k versus frequency f_k is called the *frequency spectrum* of the sampled signal x_j.

Up to now we have taken k to be all of the integers within the range of $-N/2 + 1$ to $+N/2$. However, it is easy to show that Equation(11) is periodic in k, with a period of N:

$$X_{k+N} = \sum_{j=0}^{N-1} x(t_j)\, e^{i2\pi j(k+N)/N} = \sum_{j=0}^{N-1} x(t_j)\, e^{i2\pi jk/N} e^{i2\pi j} = X_k \tag{16}$$

where we have used the fact that exp(i2πj)=1 because j is an integer. Based on Equation (16), the negative k-values in the range of $-N/2 + 1$ to -1 are equivalent to the k-value range of $+N/2 + 1$ to $+N-1$. Because the array indices in computer programs generally are positive integers, algorithms to evaluate Equation (11) typically let k vary from 0 to $N-1$ (which covers one complete period). This is the convention that LabVIEW's Fourier Transform-based VIs follow. Thus, in the N-element array of discrete Fourier transform values X_k that such a VI outputs, the elements with indices from 0 to $N/2$ contain the sequence of X_0 to $X_{N/2}$ (affiliated with positive frequencies in the range of $0 \le f \le +f_c$) and elements with indices from $N/2 + 1$ to $N - 1$ contain the sequence of $X_{-N/2 + 1}$ to X_{-1} (affiliated with negative frequencies in the range of $-f_c < f < 0$).

Fast Fourier Transform (FFT)

Since the mid-1960s, a clever algorithm to calculate the discrete Fourier transform [Equation (11)] in an efficient manner has been widely used. This algorithm, known as the *Fast Fourier Transform (FFT)*, exploits inherent symmetries in the calculation of Equation (11) that arise due to the periodic way in which the original data were taken (N evenly spaced samples taken with time-spacing Δt). Where the "brute force" evaluation of Equation (11) requires on the order of N^2 complex multiplication operations, the FFT manages this task with only $N \log_2(N)$ such operations. The savings in time are immense, especially when N is large. As an example, consider a digitizing oscilloscope configured to produce a data set of 1024 points (a typical mode of operation). To analyze the frequency spectrum of this set, the FFT algorithm will be 100 times faster than the straightforward evaluation of Equation (11). For a very large data set of 10^6 points, the FFT is 50,000 times faster!

In addition to the assumption of evenly spaced data points, there is one extra restriction to the use of FFTs. In deriving an FFT algorithm, one must assume that the size of the data set is a power of two (that is, $N = 2^m$ where m = 1, 2, 3, ...) to obtain maximum calculational efficiency. If N is not equal to a power of two, algorithms (called *Discrete Fourier Transforms* or *DFTs*) exist to evaluate the discrete Fourier transform, but with a speed much slower than that of the FFT.

FREQUENCY CALCULATOR VI

You will implement a LabVIEW FFT VI that observes the following convention: The FFT icon outputs an N-element array of discrete Fourier transform values X_k, indexed *0* to *N−1*, which are associated with N evenly spaced frequencies ($\Delta f = f_s/N$) in the range of $−f_c < f \le + f_c$. The array elements with indices from *0* to *N/2* contain the sequence of X_0 to $X_{N/2}$ associated to the positive frequencies from 0 to $+f_c$, and the elements with indices from *N/2 + 1* to *N − 1* contain the sequence of $X_{−N/2+1}$ to $X_{−1}$ associated with the negative frequencies from $(−f_c + \Delta f)$ to $−\Delta f$.

Let's first write a program, to be used later as a subVI, that calculates the frequency associated with each array element of the FFT output. Start by creating the following front panel. Assign the connector terminals consistent with the Help Window shown and design an icon. Save the VI as **Frequency Calculator** in **YourName.llb**.

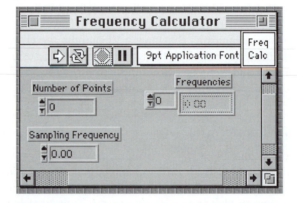

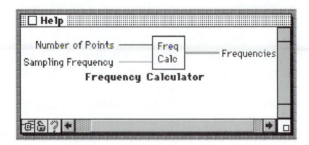

Now write a block diagram that creates an array of frequencies f_k, indexed from *0* to *N − 1*. We desire the frequency at index k to be the f_k associated with the X_k at that same index in the FFT icon's output array. The separation between adjacent frequencies f_k is $\Delta f = f_s/N$, where f_s is the sampling frequency. From the previous paragraph's discussion, we see that the f_k can be calculated as follows:

$$f_k = k\,\Delta f \qquad\qquad k = 0, 1, 2, \ldots, \frac{N}{2}$$

$$f_k = (k - N)\,\Delta f \qquad k = \frac{N}{2} + 1, \ldots, N - 1$$

(17)

Program this formula using a **Case Structure** as shown in the following illustration. An alternate approach to coding Equation (17) would be to implement the Conditional Branching function available within the Formula Node.

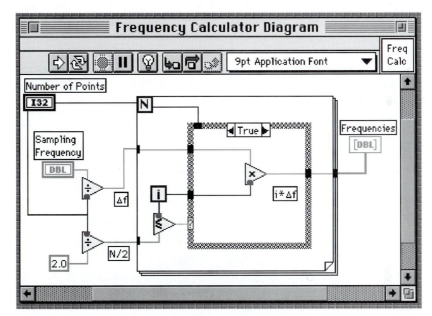

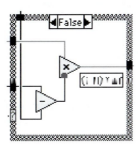

Return to the front panel. Input the values *1024* and *2000* to the **Number of Points** and **Sampling Frequency** controls, respectively. Run the VI. If it functions properly, the positive frequencies of 0 to +1000, in increments of 2000/1024 ≅ 1.95, will appear at the indices *0* through *512*. The negative frequencies of (approximately) −998.05 to −1.95 will reside at the array elements with indices from *513* through *1023*.

FFT OF SINE WAVE DATA

To get a handle on how FFT signal processing works, let's synthesize some data by summing together a few sinusoidal waves of known complex-amplitudes A_k and frequencies f_k, feed this data to the FFT algorithm, divide the resultant X_k values by N to produce the spectrum of A_k [see Equations (14) and (15)], and determine if this spectrum matches the known input. We will use our old friend **Sine Wave**, which you created in Chapter 3 and saved in **YourName.llb**, to synthesize an N-element array of data and our new friend **Frequency Calculator** to produce the array of f_k values.

Construct the front panel shown, saving it under the name **FFT of Sine Waves** in **YourName.llb**. We will display the complex-amplitudes A_k on the **XY Graph**. Since these A_k are complex numbers, we will display the real and imaginary parts of the A_k via two plots. Using the Legend, set up these two plots, labeled **Re(A)** and **Im(A)**. Distinguish these plots from each other by color.

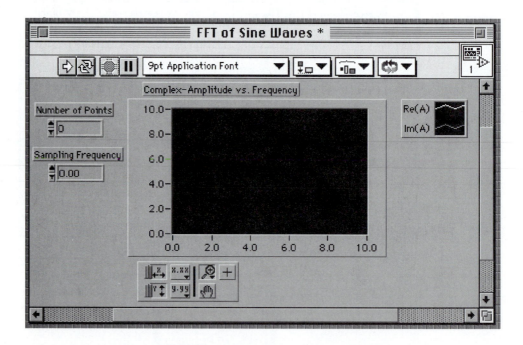

The discrete Fourier transform values X_k output by the FFT algorithm are complex numbers. LabVIEW includes a complex number data representation, so X_k is output from the FFT icon as this data-type. The Help Window for the FFT icon is shown next. Because our input data are purely real, I have selected the **Real FFT.vi** icon in **Functions>>Analysis>>Digital Signal Processing**. The more general **Complex FFT.vi**, which accepts a complex-valued input, is also available in this menu.

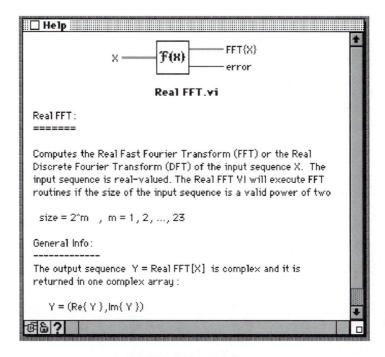

Now write the block diagram as follows. The **Complex To Re/Im** icon is found in **Functions>>Numeric>>Complex.**

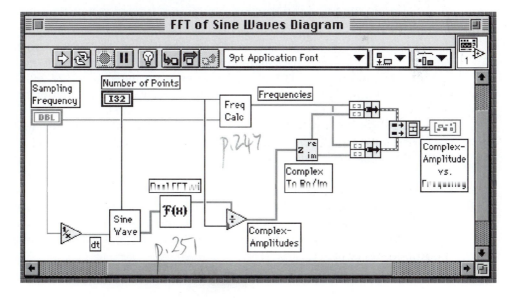

Store your final work on **FFT of Sine Waves** using **Save.**

Applying the FFT to Various Sinusoidal Inputs

Open the **Sine Wave** VI in the **YourName.llb** Library. Remember there are two easy ways to open this subVI: double click on its icon on the **FFT of Sine Waves** diagram or select **Sine Wave** in the **Unopened SubVIs** palette of the **Project** menu. Else, there's always the hard method—use the **Open** command under the **File** menu.

Switch to **Sine Wave**'s block diagram and program the Formula Node to calculate a cosine function with an *amplitude* of 4.0 and a frequency of 250 Hertz as shown. Note: I will use the word *amplitude* to mean the peak-height of a sinusoidal function.

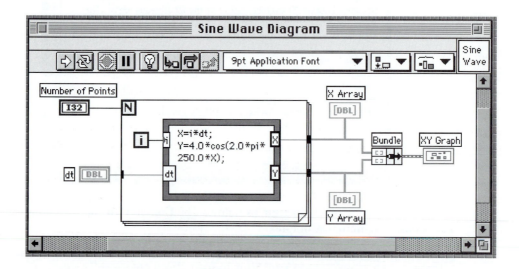

Now return to the front panel of **FFT of Sine Waves.** The **Windows** menu provides an easy way to switch back and forth between open panels and diagrams.

On the front panel of **FFT of Sine Waves**, input the values *1024* and *2000* for **Number of Points** and **Sampling Frequency**, respectively. Remember that the FFT algorithm wants the number of samples in your input data to be a power of two, so we have chosen $N = 2^{10} = 1024$. Run your program. The VI will find the real and imaginary parts of the complex-amplitude A_k for positive and negative frequencies f_k up to the Nyquist frequency $\pm f_c = \pm f_s/2$. For our choice of $f_s = 2000$ Hertz, the Nyquist frequency is 1000 Hz, so the 250 Hz cosine oscillation frequency that you have input will fall within the FFT's range of detection.

If all works well, the VI should output the Fourier transform of your cosine input in the following form: The real part of the complex-amplitude A (the y-axis quantity) is equal to zero at all frequencies f, except $A = +2$ at $f = \pm 250$. The imaginary part of A is zero at all frequencies f.

Why is this output the correct representation for the cosine input? Shouldn't the complex-amplitude A be 4, not 2? To understand your result, plug the output back into Equation (14). Then

$$x_j \approx \sum_{k=0}^{N-1} A_k\, e^{i2\pi f_k t_j} = (+2)\, e^{i2\pi(250)t_j} + (+2)\, e^{i2\pi(-250)t_j}$$

$$= 4\left[\frac{e^{i2\pi(250)t_j} + e^{i2\pi(-250)t_j}}{2}\right]$$

$$= 4\cos[2\pi(250)t_j]$$

where we have used the identity $\cos x = (e^{ix} + e^{-ix})/2$ for the last step. So we see that the output complex-amplitudes are indeed the correct representation for an input cosine function of amplitude 4.0.

Try a sine wave input and see what you get. Switch back to the block diagram of the **Sine Wave** VI and program the Formula Node to calculate a sine wave of amplitude 4.0 and frequency of 250.0 Hz.

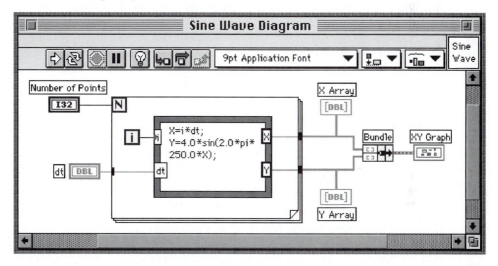

Now run **FFT of Sine Waves** with *1024* and *2000* for **Number of Points** and **Sampling Frequency**, respectively. You should find that the imaginary part of A equals -2 at $f = +250$ Hz and $+2$ at $f = -250$ Hz and is zero elsewhere. The real part of A is zero at all frequencies. To see that this is the correct representation of your input, we again start with Equation (14) and note the following:

$$x_j \approx \sum_{k=0}^{N-1} A_k\, e^{i2\pi f_k t_j} = (-2i)\, e^{i2\pi(250)t_j} + (+2i)\, e^{i2\pi(-250)t_j}$$

$$= 4\left[\frac{e^{i2\pi(250)t_j} - e^{i2\pi(-250)t_j}}{2i}\right]$$

$$= 4\sin[2\pi(250)t_j]$$

where we have used the facts that $i = -1/i$ and $\sin x = (e^{ix} - e^{-ix})/2i$.

There are two frequencies that behave differently from the rest, zero frequency and the Nyquist frequency. First, program **Sine Wave** with the constant (zero frequency) function

$$x_j = 4.0$$

then run **FFT of Sine Waves** with **Number of Points** and **Sampling Frequency** equal to *1024* and *2000*, respectively. Note that, rather than finding half of the amplitude at a positive and negative frequency as in the previously studied non-zero frequency cases, the full amplitude (called the *DC component*) appears at the single frequency f = 0.

Now, program **Sine Wave** with the following function that oscillates at the Nyquist frequency

$$x_j = 4.0 \cos[2\pi(1000.0)t_j]$$

then run **FFT of Sine Waves** again with **Number of Points** and **Sampling Frequency** equal to 1024 and 2000, respectively. If a mysterious diagonal line appears in the resulting FFT plot, remove it by popping up on the Legend and deselecting the "connect the dots" mode of plotting in the **Interpolation** palette. Similar to the zero-frequency case, you'll discover that the full oscillatory amplitude appears at the single frequency f_c.

Here is a synopsis of our last observation: The cosine function, which is used to generate the data set, takes on its peak value when the 2000 Hz sampling process begins at time t = 0. The 1000 Hz Nyquist-frequency waveform then is sampled twice each cycle, producing a data set that has the following sequence—peak value, trough value, peak value, trough value... and so on. We find that by applying the FFT algorithm to this data set, the correct value for the amplitude of oscillation (i.e., 4.0) is determined at f_c.

Unfortunately, by exploring the Nyquist-frequency situation a little further, a troublesome problem emerges. Try programming **Sine Wave** to produce a Nyquist-frequency oscillation, but this time one that follows the sine function:

$$x_j = 4.0 \sin[2\pi(1000.0)t_j]$$

then run **FFT of Sine Waves** with **Number of Points** and **Sampling Frequency** equal to *1024* and *2000*, respectively. You'll find that the FFT detects no amplitude of oscillation at f_c in this case. Can you explain why the 1000 Hz sinusoidal oscillation is invisible in this data set? If interested, you might try including a phase constant δ (in radians) in the sine function's argument

$$x_j = 4.0 \sin[2\pi(1000.0)t_j + \delta]$$

to produce a data set that neither takes on its peak value or zero value (like the cosine and sine functions, respectively) at time t = 0 . Does the FFT yield the correct amplitude of oscillation (i.e., 4.0) for any non-zero value of δ?

To summarize all of our Nyquist-frequency findings: In taking the FFT of a discretely sampled data set, the resultant value for the amplitude at f_c is only accurate when the Nyquist-frequency oscillation is "in phase" with the sampling process. Since one cannot guarantee that this will be the case in an actual experimental situation, the complex-amplitude at f_c should always be viewed with suspicion. Thus, in our work below, we will ignore the Fourier component derived at the Nyquist frequency.

Next, try programming the **Sine Wave** VI with a function that includes two simultaneous sinusoidal oscillations such as

$$x_j = 8.0 + 4.0 \sin[2\pi(250)t_j] + 6.0 \cos[2\pi(500)t_j]$$

and then run **FFT of Sine Waves.** Do you understand the output?

Finally, what happens when the input contains a sine and cosine oscillation, both at the same frequency? Program **Sine Wave** with the function

$$x_j = 4.0 \sin[2\pi(250)t_j] + 6.0 \cos[2\pi(250)t_j]$$

and then run **FFT of Sine Waves.** You may have to change, for example, the **Point Style** of each plot to view the output accurately. Do you understand the real and imaginary plots?

Magnitude of the Complex-Amplitude A_k

To this point, we have been representing a complex number as the sum of a real and an imaginary part, that is, $z = x + iy$. However, as you know, the complex number z may also be represented as a *magnitude r* times a *phase factor* $e^{i\phi}$, where $r = \sqrt{x^2 + y^2}$ and $\tan\phi = y/x$. Let's say that the oscillation of a system at frequency f is, as in the above example, the composite of a sine and cosine function with (real) amplitudes B and C, respectively:

$$x = B \sin(2\pi ft) + C \cos(2\pi ft) \tag{18}$$

Then, defining $\theta = 2\pi ft$, and using the complex exponential representation of the sine and cosine functions,

$$x = B\frac{e^{i\theta} - e^{-i\theta}}{2i} + C\frac{e^{i\theta} + e^{-i\theta}}{2}$$

$$= \frac{1}{2}(C - iB)\, e^{i\theta} + \frac{1}{2}(C + iB)\, e^{-i\theta}$$

From the following illustration, we see that $(C - iB) = re^{-i\phi}$ and $(C + iB) = re^{+i\phi}$, where $r = \sqrt{B^2 + C^2}$ and $\theta = \tan^{-1}(B/C)$.

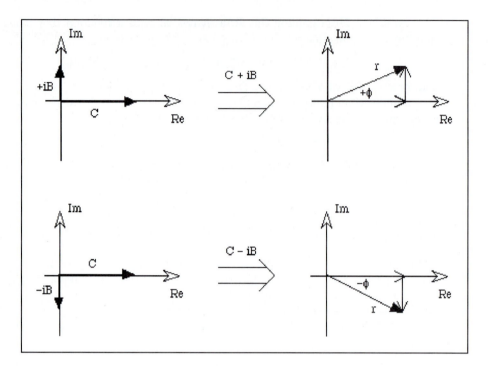

Thus,

$$x = \frac{1}{2}(r\, e^{-i\phi})\, e^{i\theta} + \frac{1}{2}(r\, e^{+i\phi})\, e^{-i\theta}$$

$$= r\, \frac{e^{i(\theta - \phi)} + e^{-i(\theta - \phi)}}{2} \tag{19}$$

$$= r \cos(\theta - \phi)$$

Then, equating (18) with (19) and remembering $\theta = 2\pi ft$, $r = \sqrt{B^2 + C^2}$, and $\phi = \tan^{-1}(B/C)$, we find that

$$x = B \sin(2\pi ft) + C \cos(2\pi ft) = \sqrt{B^2 + C^2} \cos(2\pi ft - \phi) \tag{20}$$

Equation (20) tells us that a sine and cosine oscillation at the frequency f combine to produce a single phase-shifted cosine oscillation at frequency f with a net amplitude equal to the vectorial sum of the sine and cosine amplitudes.

Let's then write a VI called **FFT–Magnitude Only** that, given an input data set x_j, finds the net amplitudes of oscillation at the frequencies f_k, but ignores the phase information (that is, neglects whether this oscillation is in the form of a pure sine wave or a pure cosine wave or a composite of sine plus cosine). With **FFT of Sine Waves** open, select **Save As…** from the **File** menu, and create a new VI named **FFT–Magnitude Only**. Use the Legend to request only a single plot—labeled **Mag(A)**—on the **XY Graph**. Relabel the **XY Graph** as **Magnitude of Complex-Amplitude vs. Frequency.**

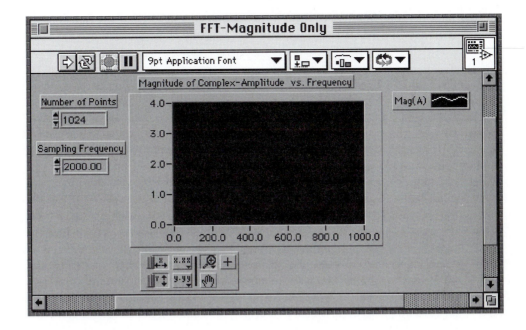

Now modify the block diagram so that, using the complex-amplitude's real and imaginary parts, its magnitude is calculated and displayed. Replace the **Complex To Re/Im** icon with **Complex To Polar** (found in **Functions>>Numeric>>Complex**). The simple way to make this swap is by popping up on **Complex To Re/Im**, then selecting the **Replace** menu item. It will be obvious what to do from that point. Complete the modification necessary to produce the block diagram shown here.

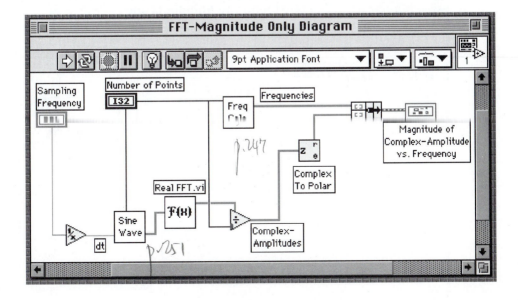

Program **Sine Wave** to calculate the sum of a sine and cosine function, both having frequency f = 250 Hz, with amplitudes 3.0 and 4.0 respectively:

$$x = 3.0 \sin[2\pi(250)t] + 4.0 \cos[2\pi(250)t]$$

From Equation (20), the above waveform should be equivalent to

$$x = \sqrt{3^2 + 4^2} \cos[2\pi(250)t - \tan^{-1}(3/4)]$$

Run **FFT–Magnitude Only** with **Number of Points** and **Sampling Frequency** equal to *1024* and *2000*, respectively. Is the output representative of a single 250 Hz cosine curve with net amplitude $\sqrt{3^2 + 4^2} = 5$?

The Fourier transform method represents a function as the linear combinations of the complex exponential basis set. Because $\cos(2\pi ft) = (e^{i2\pi ft} + e^{-i2\pi ft})/2$, a cosine function's amplitude is equally divided between the complex-amplitudes of the positive and negative frequency basis functions $e^{\pm i2\pi ft}$. Two exceptions to this "equal division of amplitude" occur, of course, for the basis functions at zero frequency (i.e., $e^{i2\pi(0)t} = 1$) and at the Nyquist frequency f_c. When displaying the spectrum of a given input data set, this symmetry of complex-amplitudes in frequency-space is commonly exploited by simply doubling the magnitude of the positive frequency's complex-amplitude and only plotting the positive-frequency axis. Such a plot displays the net amplitude of sinusoidal oscillation at each frequency $|f_k|$ directly, while phase information is completely neglected.

Modify **FFT–Magnitude Only** to plot the input data's spectrum in the manner described above. Let's ignore the Fourier component at f_c because, as we saw previously, its value is sometimes suspect. Since the y-axis values are now equivalent to the net amplitude of sinusoidal oscillation at each (positive) frequency f_k, relabel the **XY Graph** as **Amplitude vs. Frequency**. In the front-panel Legend, rename the plot as **Amplitude**. The following block diagram accomplishes the desired features.

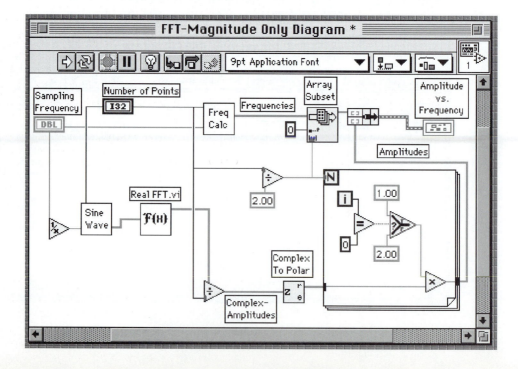

Run the VI and verify that it performs correctly. Once satisfied, save your work.

LEAKAGE AND WINDOWING

I have good news and I have bad news.

Observing Leakage

First, the bad news. Reprogram **Sine Wave** with the function

$$x = 4.0 \cos[2\pi(250.0)t]$$

and then run **FFT-Magnitude Only** using the values *1024* and *2000* for **Number of Points** and **Sampling Frequency,** respectively. Its spectrum will appear as shown.

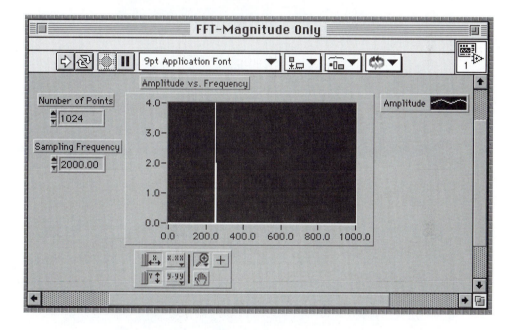

Using the Legend, make the individual plotted points visible by changing the **Point Style**. Then magnify the spectral peak at f = 250 Hz by changing the x-axis end points. An easy way to zoom in on a peak is to click on the Palette's **Magnifying Glass** and select the **X-Axis Zoom** option.

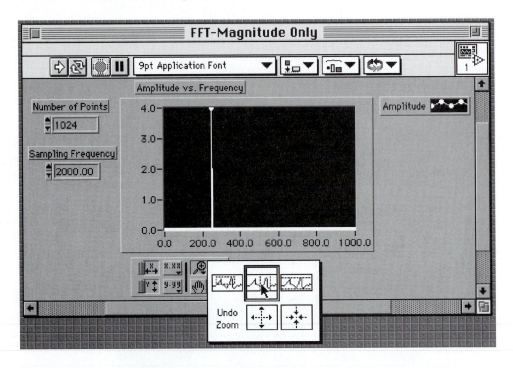

The mouse cursor will become a small magnifying glass when positioned over the plot. Click and drag it to select the region that you would like to zoom in on.

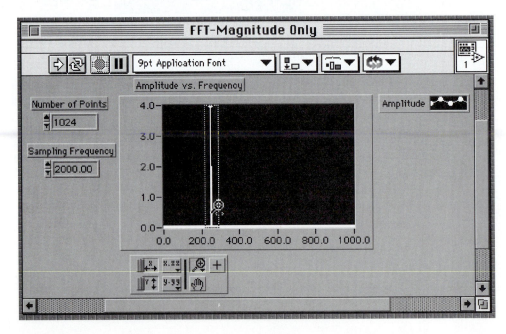

When you release the mouse button, the x-axis will be rescaled in the desired manner. You can return to the original plot by selecting **Undo Zoom** from the **Magnifying Glass** menu or by turning off, then on, the **X-Axis Autoscaling** slider switch.

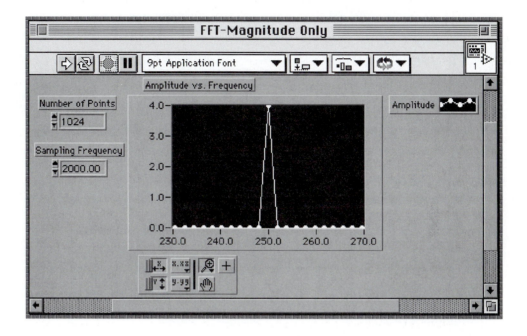

The spectrum looks perfect, doesn't it? The amplitude is equal to zero everywhere, except at the frequency of 250 Hz, where it equals +4.0, just as we expected.

So, what's the problem? Try repeating the above procedure after programming **Sine Wave** with the following function

$$x = 4.0 \cos[2\pi(249.0)t]$$

Surprisingly, the resulting spectrum no longer appears as a "delta-function" spike with a height of 4.0, but instead is a broadened peak of maximum height less than 4.0, as shown.

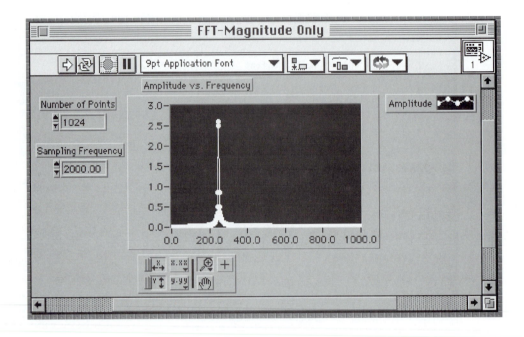

Zoom in on the peak to get a closer look.

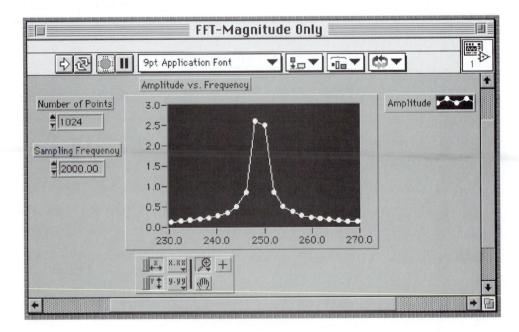

At face value, this spectrum seems to be telling us that the input data is oscillating at several different frequencies in the approximate range of 230 to 270 Hz. However, we know that this is simply untrue. We input a pure sine wave with the exact frequency of 249 Hz.

We are then faced with a mystery. Why does our FFT-based VI work perfectly for an input frequency of 250 Hz, but fails when that frequency is changed to 249 Hz? Well, here's the resolution to this paradox: I've been sneaky and always, with the exception of f = 249 Hz, asked you generate input data using sine and cosine functions that oscillate at one of the N discrete frequencies f_k. For example, via Equation (7), when acquiring 1024 equally spaced data points at a sampling rate of 2000 Hz, the FFT algorithm produces complex-amplitudes A_k at the N frequencies f_k given by

$$f_k = k\left(\frac{2000 \text{ Hz}}{1024}\right) \qquad k = -511, \ldots, 0, \ldots, 512$$

You can verify that 250 Hz and 500 Hz are both one of the f_k. In particular, they are f_{128} and f_{256}, respectively.

A way of summarizing our above observations is this: If you input a sinusoidal oscillation at exactly one of the frequencies f_k, for example $f_{128} = 250.00$ Hz or $f_{256} = 500.00$ Hz, the FFT algorithm will perfectly produce the frequency spectrum of that data (that is, a delta-function spike of the correct height at the correct frequency). However, if instead one inputs a sinusoidal oscillation at a frequency f not equal to one of the f_k, such as 249 Hz, which falls between $f_{127} = 248.05$ and $f_{128} = 250.00$, the resulting spectrum is imprecise. In fact, it is as if its spectral amplitude, which rightfully should be a spike at frequency f, has diffused from this central point and distributed itself among the neighboring f_k. This smearing of spectral information—termed *leakage*—is an artifact of the finite number N of data points contained in our discretely sampled data set. You can prove this for yourself by rerunning **FFT-Magnitude Only** (with **Sine Wave** programmed to calculate a 249.0 Hz cosine wave), first making **Number of Points** equal to *512*, then *1024*, then *2048*, followed by *4096*, etc. You will find that as N increases, the spectrum becomes much more delta-function-like.

Analytic Description of Leakage

It's not too hard to derive analytic expressions that explain these observations. Consider an oscillatory waveform with frequency f that is described by the complex exponential $x(t) = A \exp(i2\pi ft)$. Once the features of this waveform's "single spike" Fourier transform are understood, the properties of a sinusoid's "dual-spike" spectrum will be obvious. Applying the discrete Fourier transform [Equation (11)] to the complex-exponential waveform, the values of X_k are

$$X_k = \sum_{j=0}^{N-1} A \, e^{i2\pi f(j\Delta t)} \, e^{-i2\pi jk/N}$$

so

$$X_K = A \sum_{j=0}^{N-1} \left[e^{i2\pi (f\Delta t - k/N)}\right]^j \tag{21}$$

This series is the well-known finite geometric series, whose form and summation value are given by

$$\sum_{j=0}^{N-1} x^j = \frac{1 - x^N}{1 - x} \tag{22}$$

After using (22) to evaluate the sum in (21), a few lines of algebra and trigonometric relations yield the following expression for the magnitude of the discrete Fourier transform values X_k

$$|X_k| = \left| A \frac{\sin\left[\pi N \left(f\Delta t - k/N\right)\right]}{\sin\left[\pi \left(f\Delta t - k/N\right)\right]} \right| \tag{23}$$

This equation then describes the quantity determined by **FFT–Magnitude Only**.

Let's investigate the meaning of Equation (23). First, consider the case when the input frequency f exactly equals one of the frequencies f_k, say $f_{k'}$. Setting $f = f_{k'} = k'$ (f_s/N) = k' ($1/N\Delta t$), Equation (23) becomes

$$|X_k| = \left| A \frac{\sin[\pi (k' - k)]}{\sin[\pi (k' - k)/N]} \right| \tag{24}$$

For $k \neq k'$, the difference $(k' - k)$ will be an integer less than N, making the numerator and denominator of (24) zero and non-zero, respectively. Thus, $|X_k| = 0$ for $k \neq k'$. However, for $k = k'$, both numerator and denominator of (24) are zero and l'Hopital's rule gives $|X_{k'}| = A N$. So we expect the resulting frequency spectrum to be a delta function with a single spike at frequency $f_{k'}$ of height $|A_{k'}| = |X_{k'}|/N = A$. This prediction is consistent with what we observed when we input the oscillations at $f = f_{128} = 250$ Hz and $f = f_{256} = 500$ Hz inputs to the FFT algorithm.

Second, consider the case when the input frequency f is not equal to one of the frequencies f_k. We can simplify the appearance of Equation (23) by noting

$$f\Delta t - k/N = \frac{f}{f_s} - \frac{f_k}{f_s} = \frac{f - f_k}{f_s}$$

Then Equation (23) becomes

$$|X_k| = \left| A \frac{\sin\left[\pi N \left(\dfrac{f - f_k}{f_s}\right)\right]}{\sin\left[\pi \left(\dfrac{f - f_k}{f_s}\right)\right]} \right|$$

and, since $\Delta f = f_s/N$,

$$|X_k| = \left| A \frac{\sin\left[\pi \left(\dfrac{f - f_k}{\Delta f}\right)\right]}{\sin\left[\pi \left(\dfrac{f - f_k}{f_s}\right)\right]} \right| \tag{25}$$

Inspecting Equation (25), we see that, under our assumption that f does not equal one of the f_k, none of the $|X_k|$ will be zero. However, the frequencies f_k that fall closest to f will make the denominator of Equation (25) the smallest. Thus the $|X_k|$ associated with the f_k neighboring f will take on the largest values. To better visualize the meaning of Equation (25), I use it to plot the magnitude of the complex-amplitude $|A_k| = |X_k| /N$ versus f_k with the following familiar choice of parameters: $f_s = 2000$ Hz, $N = 1024$, A = 4.0, and f = 249 Hz. Then $\Delta f = f_s/N = 2000/1024$ and $f_k = \Delta f\,k$, where k = –512, ..., 0, ..., 511.

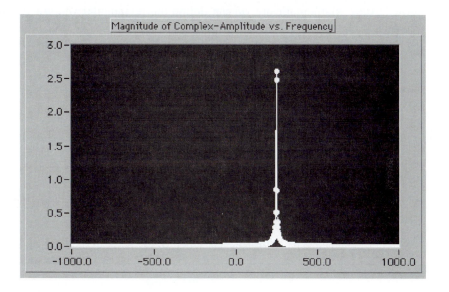

Zooming in on the peak to get a closer look, we see that Equation (25) provides a perfect theoretical prediction of the spectral leakage we observed previously with a 249 Hz input to **FFT-Magnitude Only**.

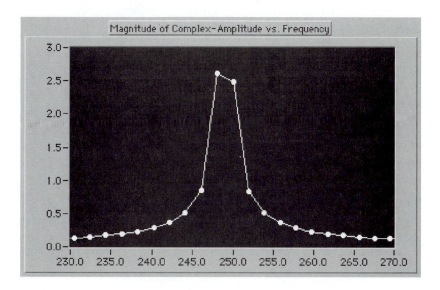

Description of Leakage Using the Convolution Theorem

The *Convolution Theorem*, a powerful result from higher mathematics, provides another vantage point from which to understand the problem of leakage. From this point of view, the situation appears as follows: When we acquire a finite number N of discretely sampled points for FFT spectral evaluation, we are in effect observing an infinite set of data d_j ($j = -\infty, ..., -1, 0, +1, ..., +\infty$) through a rectangular viewing window in time. Mathematically, we can define the rectangular window function $w(t_j)$ to be zero at all times $t_j = j\,\Delta t$, except during the "data-viewing" time interval from $j = 0$ to $j = N - 1$, when it is equal to one. Then our finite set of N sampled data points x_j is given by the product $x_j = d_j\,w_j$. This idea is illustrated here.

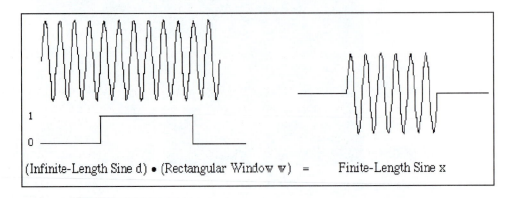

(Infinite-Length Sine d) • (Rectangular Window w)   =   Finite-Length Sine x

Let's say that the Fourier transforms of d_j and w_j are D_k and W_k, respectively. The question relevant to our finite-length data set then becomes, "What happens when one takes the Fourier transform of the product $d_j w_j$ ($= x_j$)?" Well, according to the famous Convolution Theorem, the Fourier transform of the product of two functions $d_j\,w_j$ is equal to the *convolution* of the two Fourier transforms D_k and W_k. The convolution for continuous functions, denoted by D*W, is defined by

$$D(f)*W(f) = \int_{-\infty}^{+\infty} D(\phi)W(f - \phi)\,d\phi \tag{26}$$

In the discrete case, this definition becomes

$$(D*W)_k = \sum_{m=-N/2+1}^{N/2} D_m\,W_{k-m} \tag{27}$$

The convolution of two functions, while complicated in general to determine, is simple to ascertain in the following important case. Let D_m be a unit-amplitude delta function located at the frequency f_n, that is, $D_m = 0$ for all m, except $D_{m=n} = 1$. Then, from (27), we find that $D*W = W_{k-n}$, meaning that the convolution is just the function W displaced so that it is now centered at f_n rather than $f = 0$. This idea is illustrated below using continuous functions.

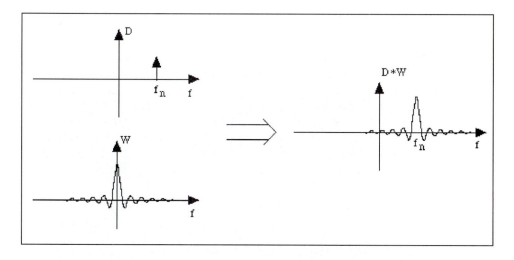

This example provides new insight into the leakage phenomenon. Consider the case of a finite-length complex-exponential input of the form $x(t_j) = A \exp(i2\pi ft_j)$, which can be described as the product of an infinite-length complex-exponential $d(t_j)$ and a rectangular window function $w(t_j)$. The Fourier transform D of the infinite complex-exponential is, of course, a delta function of height A located at frequency $+f$, while the discrete Fourier transform of the rectangular window (apart from a unity-amplitude phase factor) is easily shown to be

$$(W_{rectangle})_k = \frac{\sin\left[\dfrac{\pi f_k}{\Delta f}\right]}{\sin\left[\dfrac{\pi f_k}{f_s}\right]} \tag{28}$$

A plot of Equation (28) (treating frequency as a continuous variable) with $f_s = 2000$ Hz and $N = 1024$ is shown in the following diagram. Note the substantial amplitudes of $W_{rectangle}$ at high frequencies, which we know qualitatively result from the sharp turn-on and turn-off at the edges of the rectangular window (recall from the Fourier analysis of a square wave, its sharp edges are produced by the presence of high-frequency harmonics). The values of $W_{rectangle}$ at the frequencies f_k, that is, $(W_{rectangle})_k$, are given by the dots.

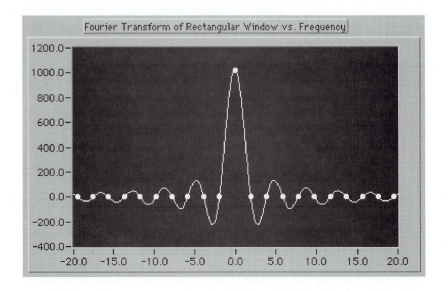

The magnitude of the finite-length complex-exponential's Fourier transform X is formed from the convolution of $W_{rectangle}$ with the height-A delta function at $+f$. This convolution is simply $W_{rectangle}$, shifted along the frequency axis so that it is centered at frequency f, then multiplied by A. Thus,

$$|X_k| = |(D*W)_k| = \left| A \frac{\sin\left[\dfrac{\pi (f_k - f)}{\Delta f} \right]}{\sin\left[\dfrac{\pi (f_k - f)}{f_s} \right]} \right| \tag{29}$$

Note that (29) is equivalent to the "leakage description" given by Equation (25).

Next, I plot Equation (29) for $f_s = 2000$ Hz and $N = 1024$ with f equal to one of the f_k, namely, $f = f_{128} = 250$ Hz. I take $A = 1$ for simplicity. Note that all of the values of

$|X_k|$, which are given by the dots, fall on the zero-crossing of Equation (29), except $|X_{128}|$ at $f_{128} = 250$ Hz. Thus the discrete Fourier spectrum will be a delta function in this case.

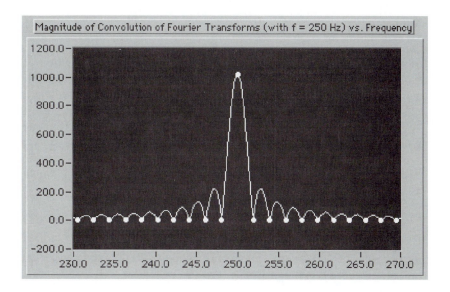

Changing f to the non-f_k value of 249 Hz, the following plot of Equation (29) is obtained. Note that the $|X_k|$, again indicated by the dots, now fall at non-zero locations on the curve defined by Equation (29), resulting in the spectral leakage phenomenon.

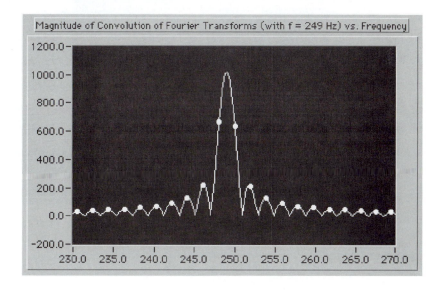

The Convolution Theorem then gives us the following insight: It is the wide-ranging frequency content of the rectangular window function (as given in Equation (28) and shown in the previous plot) that produces leakage of spectral amplitude to frequencies f_k far removed from the input frequency f.

Windowing

Now the good news. Thanks to the power of your high-speed computer, you have the option of windowing your finite-length data set with some function other than a rectangle. Once the rectangular-windowed data set $(x_{rectangle})_j$ has been collected into your computer, all you have to do is construct a more desirable window w_j in software and then form the product $x_j = w_j (x_{rectangle})_j$. Leakage can simply be suppressed by choosing a window whose Fourier transform has minimal high-frequency components. Qualitatively, we know that the leakage-producing high-frequency components result from the discontinuous jumps at the edges of the rectangular function, so the "desirable" software window functions should be constructed to have a much smoother turn-on and turn-off. The Hanning window, which is defined to be

$$(w_{Hanning})_j = 0.5\left[1 - \cos\left(\frac{2\pi j}{N}\right)\right] \qquad j = 0, 1, \ldots \ldots, N - 1$$

is a popular choice for such a software window. The Hanning and Rectangle windows are plotted here for comparison.

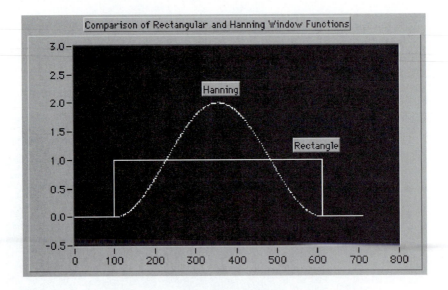

Let's try windowing to our data set in **FFT–Magnitude Only** and see how this improves things. To window the **Sine Wave**-produced data set, you will use **Scaled**

Time Domain Window.vi in **Functions>>Analysis>>Measurement**. Its Help Window is shown here.

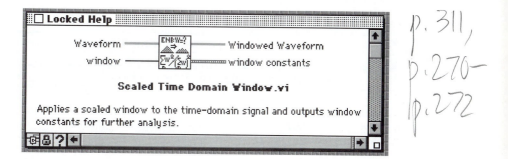

p. 311,
p. 270—
p. 272

For this icon, you provide the "to-be-windowed" data array at the **Waveform** input and select the desired window at **window** using the following integer code:

Code	Window
0	No Window
1	Hanning
2	Hamming
3	Blackman–Harris
4	Exact Blackman
5	Blackman
6	Flat Top
7	Four-Term Blackman–Harris
8	Seven-Term Blackman–Harris

The resultant array is output at **Windowed Waveform**. The cluster **window constants** is a two-element bundle of the selected window's *equivalent noise bandwidth (ENBW)* and *coherent gain*. These window parameters are important for use in some calculations, as will be shown in a few minutes.

Switch to the block diagram of **FFT–Magnitude Only**. Include **Scaled Time Domain Window.vi** as shown, so that the data set obtained from **Sine Wave** is windowed prior to being passed to the FFT icon.

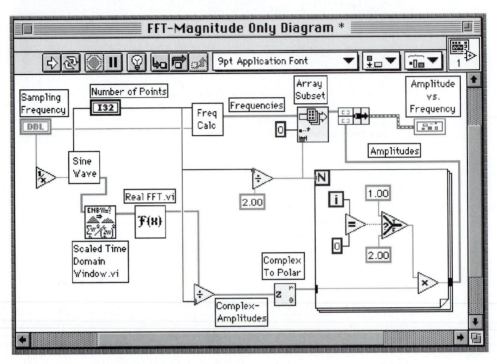

p.311

Pop up on the **window** input and select **Create Control**.

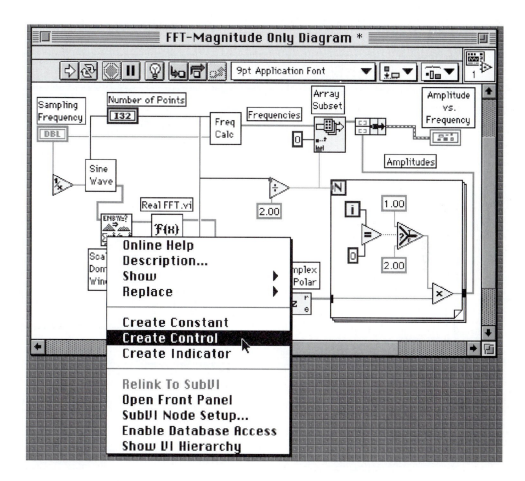

The Enumerated Type Control

The terminal for a front-panel control will then magically appear on your block dia gram, wired to the **window** input. This block diagram terminal is associated with a "*ring*" style control called (I'm still in the process of forgiving the LabVIEW developers for this) the **Enumerated Type**.

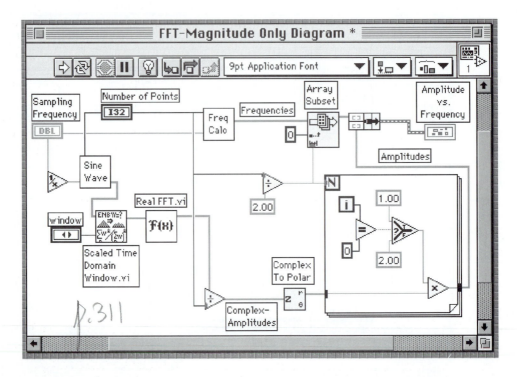

Switch to your front panel, find the **Enumerated Type** control labeled **window** that now appears there, then position it aesthetically.

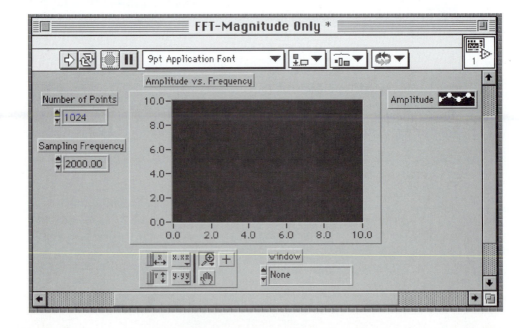

Ring and List controls—you'll find many types in **Controls>>List & Ring**—are useful when the programmer needs to present the user with a catalog of options. The **Enumerated Type** is a special type of ring control that links a sequence of text messages with a series of associated integers. When the user selects a particular text message on the front panel, its associated integer is passed to a terminal on the block diagram. In our present circumstance, the **Enumerated Type** control will provide a handy method for selecting a specific window from among the available options.

One of the nice benefits you derive from having implemented **Create Control** to produce the front-panel **window** control is that, while creating it, LabVIEW automatically loaded this **Enumerated Type** with the sequence of string messages that label the available options at Scaled Time Domain Window.vi's **window** input. To make visible the integer associated with each of these text messages, pop up on the **Enumerated Type** and select **Digital Display** in the **Show** palette.

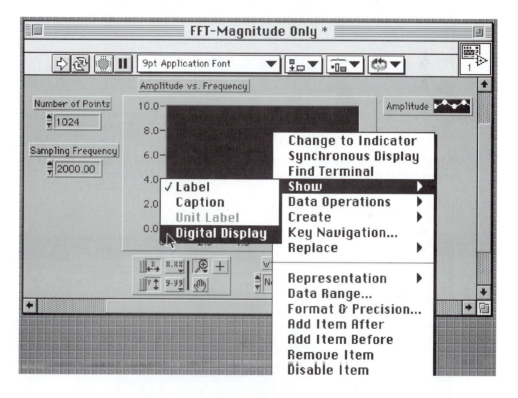

Using the Operating Tool, review the sequence of selections offered by the **window** control and verify that it is consistent with Scaled Time Domain Window.vi's **window** code given in the table above.

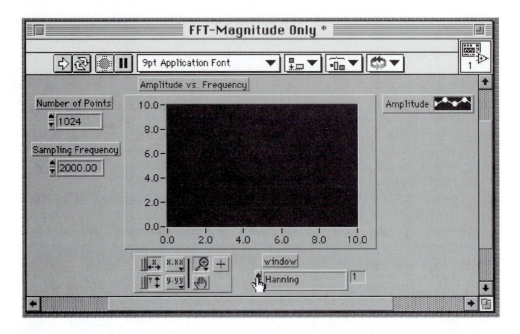

You may now wish to turn off the Enumerated Type's **Digital Display**. Save your work.

Let's explore the positive effects of windowing. Let **Number of Points** and **Sampling Frequency** be *1024* and *2000*, respectively, and program **Sine Wave** to calculate *4.0 sin[2π(249.0 Hz)t]*. Zoom in on the expected peak by scaling the axes appropriately, then turn off the x and y autoscaling options. Run the VI without the benefit of windowing by selecting *None* in the **window** control.

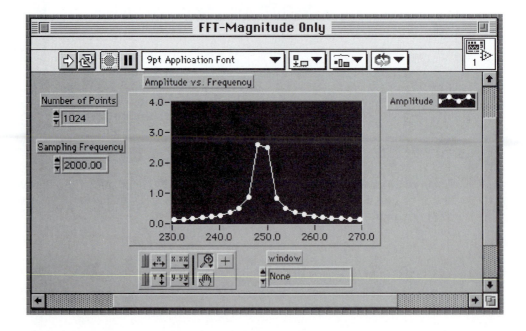

Now select the *Hanning* window and rerun the VI.

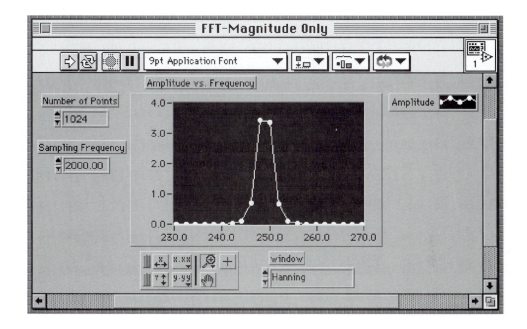

Note the dramatic decrease in leakage through the use of a Hanning window. See, I told you I had good news! Explore the result of using the other available windows.

Estimating Frequency and Amplitude

I have even more good news. Define the *power* at frequency f_k to be $P_k \equiv A_k^2$. Then the sinusoidal input frequency f that generated the finite-width spectral peaks observed above can be estimated by a weighted sum, where each observed f_k is weighted by the power at that frequency. That is,

$$ f \approx \frac{\sum_k f_k P_k}{\sum_k P_k} = \frac{\sum_k f_k A_k^2}{\sum_k A_k^2} \tag{30} $$

where the sum is over the k-values that span the peak.

Also the power $P \equiv A^2$ of the amplitude A sinusoid can be estimated by

$$ P \approx \frac{\sum_k P_k}{ENBW} = \frac{\sum_k A_k^2}{ENBW} \tag{31} $$

where *ENBW* is the effective noise bandwidth of the window used in the analysis process that produced the peak.

Write a VI called **Estimated Frequency and Amplitude** that implements Equations (30) and (31). The front panel should look something like this:

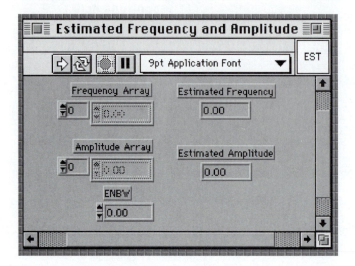

Assign the connector terminals consistent with the following Help Window.

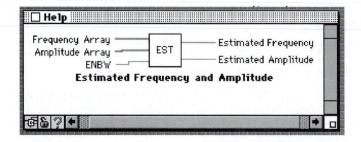

Now code the following block diagram. Because it is easy to do, we will perform the sums in Equations (30) and (31) over all k, but it is only necessary to include k-values for which the A_k are significantly non-zero, that is, k in the neighborhood of the peak.

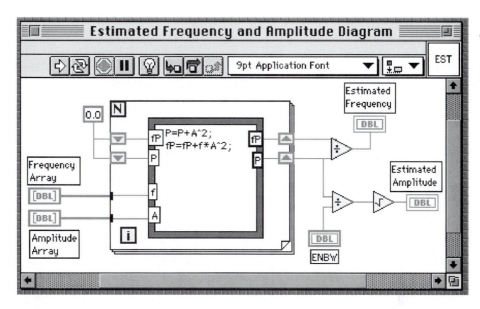

p.311

Save this VI as you close it.

Add two **Digital Indicators** (each with a large number of **Digits of Precision**) to the front panel of **FFT–Magnitude Only** and label them **Estimated Frequency** and **Estimated Amplitude.**

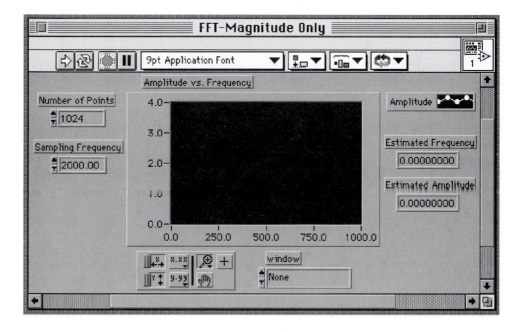

Switch to the block diagram and incorporate **Estimated Frequency and Amplitude** as shown. Unbundle (the **Unbundle** icon is found in **Functions>> Cluster**) the **window constants** cluster that is output from **Scaled Time Domain Window.vi** to obtain the selected window's **ENBW. ENBW** is the top element in the cluster, **coherent gain** (which we will not use) is the bottom element. Save your work.

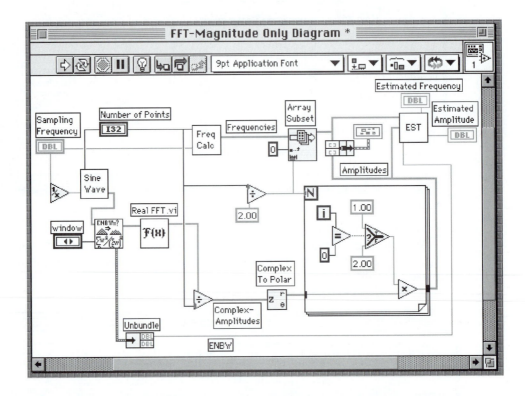

Return to the front panel and run the VI using various windows. I think you'll be very pleased with the values of **Estimated Frequency** and **Estimated Amplitude**! Can you make any qualitative conclusions about the strengths and weaknesses of the different windows?

Analog-to-Digital Conversion

Before You Come To Lab

A constant current source circuit is shown below. Use the "golden rules" of op-amps—
the voltage difference between the two inputs is zero and *no current flows into the
inputs*—to explain why the given values of R1, R2, and R3 will result in 0.1 mA of current through the thermistor. This current is small enough to prevent *self-heating* of the
thermistor, yet large enough to provide a measurable voltage across the 10 kΩ thermistor. What change(s) would you make to the circuit in order to obtain a constant current
of 0.2 mA through the thermistor?

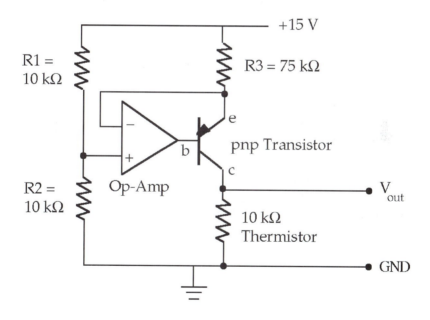

LabVIEW was developed to facilitate computer-controlled data acquisition
(DAQ) and analysis. In this chapter, you will at last learn how LabVIEW allows the
physical investigator to reach out through his or her computer and monitor events in
the outside world.

Data Acquisition VIs

LabVIEW's built-in DAQ VIs are written at three levels: *Easy I/O*, *Intermediate*, and *Advanced*. An example of how such VIs are arranged in a DAQ palette can be found in **Functions>>Data Acquisition>>Analog Input**. As is common for all DAQ palettes, the Easy I/O VIs occupy the top row, Intermediate VIs fill the middle row and a Utility subpalette, while Advanced VIs are found in their own subpalette in the lower right-hand corner.

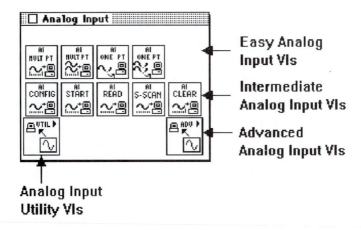

The Easy I/O VIs perform the most basic Input/Output operations such as reading a voltage level or outputting an analog signal. They are very straightforward to use and perform error checking automatically, but have limited capabilities. If your data acquisition operations are simple, you may be able to use these VIs exclusively.

At the other extreme, the Advanced VIs are the building blocks from which all other DAQ VIs are formed. They offer full control over every parameter involved in the data acquisition process and so provide the ultimate in flexibility and power to the user. This control, of course, comes at the price of added complexity to the programming of your block diagram.

Finally, the Intermediate VIs are commonly used configurations of the Advanced VIs. The Intermediate VIs offer most of the power and flexibility of the Advanced VIs, but, through the elimination of some of the more esoteric capabilities, allow the user-friendliness of fewer programming parameters.

Examine the contents of the various data acquisition palettes found in **Functions>>Data Acquisition**. There you will find built-in VIs that control all of the data-taking capabilities of a National Instruments DAQ board, once that board is present in one of the expansion slots of your computer. These capabilities include analog-to-digital (A/D), digital-to-analog (D/A), event timing, pulse counting, and digital input/output (DIO) operations. In this chapter, you will learn how to perform LabVIEW-based A/D operations.

To implement the analog-to-digital capability of your National Instruments data acquisition board, LabVIEW provides eight software building blocks—the Advanced Analog Input (AI) VIs. Each of these VIs executes one of the several steps necessary in performing the complete A/D process. In such an operation, first you must alert your computer to the presence of the DAQ board itself and its location within the available expansion slots. Then, if you wish to digitize and store an extended waveform (or, possibly, simultaneously digitize and store multiple waveforms on several input channels of your DAQ board), you need to set the rate at which data points are to be sampled and also allocate the required block of memory within your computer for storage. Finally, you must start the acquisition process, transfer the acquired data to memory, and terminate the acquisition. In the following exercises, we will identify which one of the eight AI Advanced VIs does each of the steps described.

DAQ Hardware and the Method of Successive Approximations

You might be wondering how the hardware on a National Instruments DAQ board actually accomplishes the A/D conversions. Well, it accomplishes this feat by the method of *successive approximations*. In this method, only n cycles of a clock are required to encode an analog voltage level into a number with n bits of resolution, so the conversion rate can be very fast. For example, the popular AT-MIO-16E-10, Lab-PC, and PCI-1200 boards require just 10 μsec per 12-bit A/D operation, allowing data to be sampled at a rate of up to 100,000 Hz.

The method of successive approximations, as illustrated below, makes use of a digital-to-analog (D/A) converter, which outputs a unique analog voltage V_{DA} for each possible combination of binary values at its n input bits. Then, in a process that resembles weighing an unknown object on a chemical balance using a set of various-sized weights, the analog input voltage V_{in} is compared with a particular sequence of V_{DA} values. First, a logic circuit sets the most-significant bit (MSB) at the D/A input, causing the D/A converter to output one-half of its full-scale value. The logic circuit then checks the comparator output to find if V_{in} exceeds this value and, based on this comparison, decides whether or not to keep the MSB set. In the next clock cycle, the next-most-significant bit is set and the new D/A output is compared to V_{in}. If, for example, the new value of V_{DA} is larger than V_{in}, then the comparator output goes to its low state, signaling the logic circuit to switch the current "bit-under-test" off. This process continues n times until all of the D/A converter's input bits, down to the least-significant bit (LSB), have been set, tested, and consequently kept on or switched off. Once the appropriate pattern of bits has been determined, this pattern is output as an n-bit word.

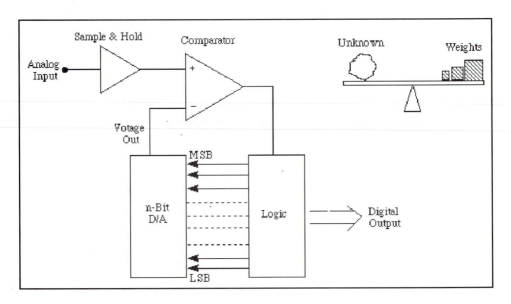

A conduit carrying signals to and from the outside world must somehow be connected to the DAQ board. For example, among its various capabilities, a PCI-1200 board has eight AI channels, each standing ready to digitize an analog input signal. So how do we get the meaning-laden analog signals from our experiment to the board's channel inputs? This access comes through a ribbon-cable connector at the end of the board with (depending on the particular board) 50, 68, or 100 pins. Each pin carries a signal corresponding to a particular function of the interface board and these functions are denoted on the pinout diagram in the board's user's manual. The pinout diagrams for the widely used E Series MIO-16, PCI-1200, and Lab-PC interface boards are reproduced in the next illustration.

I/O Connectors

E Series MIO-16

```
ACH8  — 34  68 — ACH0
ACH1  — 33  67 — AIGND
AIGND — 32  66 — ACH9
ACH10 — 31  65 — ACH2
ACH3  — 30  64 — AIGND
AIGND — 29  63 — ACH11
ACH4  — 28  62 — AISENSE
AIGND — 27  61 — ACH12
ACH13 — 26  60 — ACH5
ACH6  — 25  59 — AIGND
AIGND — 24  58 — ACH14
ACH15 — 23  57 — ACH7
DACOOUT — 22  56 — AIGND
DAC1OUT — 21  55 — AOGND
EXTREF — 20  54 — AOGND
DIO4  — 19  53 — DGND
DGND  — 18  52 — DIO0
DIO1  — 17  51 — DIO5
DIO6  — 16  50 — DGND
DGND  — 15  49 — DIO2
5 V   — 14  48 — DIO7
DGND  — 13  47 — DIO3
DGND  — 12  46 — SCANCLK
PFI0/TRIG1 — 11  45 — EXTSTROBE*
PFI1/TRIG2 — 10  44 — DGND
DGND  —  9  43 — PFI2/CONVERT*
5 V   —  8  42 — PFI3/GPCTR1_SOURCE
DGND  —  7  41 — PFI4/GPCTR1_GATE
PFI5/UPDATE* —  6  40 — GPCTR1_OUT
PFI6/WFTRIG  —  5  39 — DGND
DGND  —  4  38 — PFI7/STARTSCAN
PFI9/GPCTR0_GATE —  3  37 — PFI8/GPCTR0_SOURCE
GPCTR0_OUT —  2  36 — DGND
FREQ_OUT —  1  35 — DGND
```

PCI-1200
Lab-PC

```
ACH0          | 1  2 | ACH1
ACH2          | 3  4 | ACH3
ACH4          | 5  6 | ACH5
ACH6          | 7  8 | ACH7
AISENSE/AIGND | 9 10 | DACOOUT[1]
AGND          |11 12 | DAC1OUT[1]
DGND          |13 14 | PA0
PA1           |15 16 | PA2
PA3           |17 18 | PA4
PA5           |19 20 | PA6
PA7           |21 22 | PB0
PB1           |23 24 | PB2
PB3           |25 26 | PB4
PB5           |27 28 | PB6
PB7           |29 30 | PC0
PC1           |31 32 | PC2
PC3           |33 34 | PC4
PC5           |35 36 | PC6
PC7           |37 38 | EXTTRIG
EXTUPDATE[1]  |39 40 | EXTCONV*
OUTB0         |41 42 | GATB0
OUTB1         |43 44 | GATB1
CLKB1         |45 46 | OUTB2
GATB2         |47 48 | CLKB2
+5 V          |49 50 | DGND
```

Analog Input Modes

The above diagram can be deciphered by noting that the family of E Series MIO-16 boards have 16 analog input channels, while the PCI-1200 and Lab-PC models have 8. Through software settings (using the **NI-DAQ Configuration Utility** or LabVIEW's **AI Hardware Config.vi** icon), these inputs can be configured to operate in two distinct analog input modes: *single-ended* and *differential*. On older model DAQ boards, the input mode is selected by the proper placement of hardware jumpers on the board.

In the single-ended mode, each available analog input pin is an *AI channel*, with all of these channels referenced to the same common ground. You may choose this common ground to be the building ground (termed the *"referenced singled-ended"* mode) or supply your own ground level at the AISENSE pin (*"nonreferenced singled-ended"* mode). As an example, when operated in the referenced single-ended input mode, the pin assignments for the AI channels on the boards cited above are given in the following table. The default input mode for the PCI-1200 and Lab-PC boards is referenced single-ended.

Channel Assignments for Referenced Single-Ended Input Mode																
Channel	0	1	2	3	4	5	6	7	8	9	10	11	12	13	14	15
E Series MIO pin (GND is pin 67, 32, 64, 29, 27, 59, 24, 56)	68	33	65	30	28	60	25	57	34	66	31	63	61	26	58	23
PCI-1200, Lab-PC pin (GND is pin 9)	1	2	3	4	5	6	7	8								

In differential input mode, available analog-input pins are paired to form eight (E Series MIO-16) or four (PCI-1200, Lab-PC) independent AI channels, each of which is sensitive only to the voltage difference between paired pins. In this "differential amplifier" configuration, noise pickup by the AI channels is suppressed. Thus, if it is not a problem in your situation to halve the number of available measurement channels, the differential input mode is the preferred *modus operandi*. By default, the E Series MIO boards are configured in the differential input mode. Again, as an example, the pin assignments for the differential input mode AI channels for our representative boards are given in this table.

Channel Assignments for Differential Input Mode			
Interface Board	Channel Number	Positive Input Pin	Negative Input Pin
E Series MIO-16	0	68	34
	1	33	66
	2	65	31
	3	30	63
	4	28	61
	5	60	26
	6	25	58
	7	57	23
PCI-1200 Lab-PC	0	1	2
	1	3	4
	2	5	6
	3	7	8

An I/O Connector Block (with screw terminals to which wires from your experiment can be attached) and a ribbon cable offer a handy method of linking your experiment to the board's I/O Connector. Alternately, an adapter board with BNC connectors and a ribbon cable can be used. These items may be purchased from National Instruments.

Range, Gain and Resolution

When measuring analog signals with a DAQ board, there are several issues to keep in mind. First, the board can only measure voltages that fall within an accepted *range*. A

National Instruments DAQ board typically offers the choice of several ranges (such as 0V to +10V, −10V to +10V, and −5V to +5V) with the choice being made by a software setting in **NI-DAQ Configuration Utility** or LabVIEW's **AI Hardware Config.vi**. On older model boards, the choice is made by hardware jumpers. Check the spec sheet for your particular DAQ board to find the available ranges and method of selection. Next, so that you can match your signal range to that of your board's A/D converter, the board includes a pre-amp that can amplify the analog signal with a selectable *gain* G, prior to being digitized. For example, the available values for G on the PCI-1200 board are 1, 2, 5, 10, 20, 50 and 100 with the choice being software-programmable. Again, check the spec sheet for your particular board to find the available values of gain. Finally, the board's A/D converter represents the analog voltage level being sampled with a binary number of n-bits, which places a limit on the smallest detectable voltage difference ΔV. That is, the converter divides the measurable voltage span V_s ($V_s = 20$ Volts when the measurable range is from −10 V to +10 V) into 2^n divisions, and so the resulting *resolution* of the original analog input signal will be

$$\Delta V = \frac{V_s/G}{2^n} \tag{1}$$

For the typical values of $V_s = 20$ Volts and $G = 1$, the resolution provided by a 12-bit and 16-bit board is 5 mV and 0.3 mV, respectively.

Sampling Frequency and the Aliasing Effect

Another important issue peculiar to digitized data is related to the sampling rate f_s, where f_s is the frequency with which A/D conversions take place. The sampling frequency places an upper limit on the range of frequencies allowed within the original analog signal, if the digitizing process is to result in a faithful representation of the input. In the previous chapter, we learned that this upper threshold is called the Nyquist frequency f_c ($=1/2\ f_s$). The physical significance of the Nyquist frequency is this: For a given sampling rate, an input sine wave of frequency f_c will be sampled in the minimal manner necessary to represent the sinusoid's peaks and valleys—just twice per cycle. If the input frequency exceeds this threshold, the sampling rate becomes insufficient and the A/D conversion process becomes inaccurate in a way described next.

Fairly often, the bandwidth limitation placed on a "to-be-digitized" input analog signal does not cause you—the experimentalist—much of a problem. For example, if dealing with audio-produced electrical signals, you might know on physical grounds that the frequency content of the analog input is bracketed by zero and 20,000 Hz. Then, with a sampling rate of at least 40 kHz, this input can be properly acquired. Alternately, an analog signal may have passed through an amplifier, which behaves as a low-pass filter due to its finite bandwidth response. In this case, the sampling rate must be twice the maximum frequency passed by the amplifier. If no natural frequency bracketing exists in your experiment, then you must impose a high-frequency cutoff by placing a low-pass filter in your data-gathering circuitry. Given the available sampling rate, the filter's components then are chosen so that frequencies higher than $f_s/2$ are not passed.

What happens if a frequency higher than the Nyquist limit accidentally is input to your digitizer? Something much worse than you might expect. In a process called *aliasing*, that too-high frequency appears falsely as a lower-than-Nyquist frequency when processed by the digitizer. This phenomenon, which is peculiar to discrete sampling, is illustrated here.

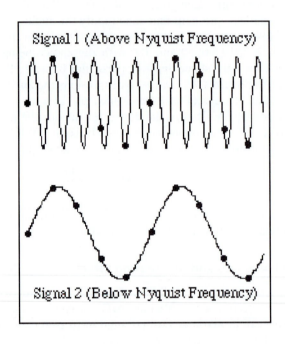

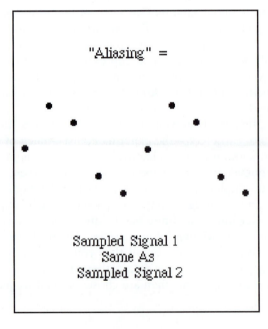

Quantitatively, we note that, if an input signal is digitized at times $t = j\Delta t$ ($j = 0, 1, 2, ...,$ $N-1$), two waves given by $\exp[i2\pi f_1 t]$ and $\exp[i2\pi f_2 t]$ will appear equal if

$$f_1 - f_2 = n\, f_s \qquad (2)$$

where n is an integer. To prove this assertion, note the following:

$$e^{i2\pi(f - nf_s)t} = e^{-i2\pi nf_s(j\Delta t)}\, e^{i2\pi ft} = e^{-i2\pi nj}\, e^{i2\pi ft} = 1\, e^{i2\pi ft}$$

where we have used the fact that $f_s \Delta t = 1$ and that the product of two integers *nj* is itself an integer. Let one of the frequencies on the left-hand side of Equation (2) be the higher-than-Nyquist input frequencey f_{input} and the other be the lower-than-Nyquist alias frequency f_{alias}. When using a VI (such as **FFT-Magnitude Only**) which determines the input sinusoid's amplitude, while ignoring its phase, Equation (2) can be rewritten as follows:

$$f_{alias} = |f_{input} - nf_s| \qquad \text{(Aliasing Condition)} \qquad (3)$$

where the integer n is chosen so that $0 \leq f_{alias} \leq f_c$.

Demonstrate the aliasing effect by opening **FFT-Magnitude Only** and choosing **Number of Points** and **Sampling Frequency** to be *1024* and *2000*, respectively. Run this VI with the subVI **Sine Wave** programmed to calculate *4.0cos[2π(250.0 Hz)t]* to remind yourself what the FFT of a sinusoid with frequency less than the Nyquist frequency (in this case, $f_c = 1000$ Hz) looks like. Now program **Sine Wave** to calculate *4.0cos[2π(2250.0 Hz)t]* and rerun **FFT-Magnitude Only**. You should find that this "higher-than-Nyquist" frequency wrongly appears in the FFT spectrum as a frequency of |2250 Hz − 2000 Hz| = 250 Hz, as given by Equation (3) with n = 1. Predict what will happen if you program **Sine Wave** with some other "higher-than-Nyquist" frequencies such as 1250 Hz and 3500 Hz, then see if your expectations are realized via use of **FFT-Magnitude Only.**

I'll assume that you have a National Instruments data acquisition board plugged into one of your computer's expansion slots and that its driver software has been correctly installed. With these tasks completed, the only remaining chore is to determine the *device number* for your board. With newer versions of LabVIEW, you can determine the device number for your DAQ board by using the **DAQ Wizards** found in the **Project** menu or by running a program called **NI-DAQ Configuration Utility.** For Windows 3.1 users, run **WDAQCONF**; on an older Macintosh system, open **NI–DAQ** in the **Control Panels** folder.

SIMPLE ANALOG-TO-DIGITAL OPERATION ON A DC VOLTAGE

Let's build a VI that, by controlling simple analog-to-digital operations, will turn your computer into a voltmeter. Construct the following front panel. If, through Equation (1), you know the precision ΔV of the digitizer on your National Instruments board, use this information to select the proper number of **Digits of Precision** on the **Voltage** indicator. For example, if $\Delta V = 0.3$ mV, **Digits of Precision** should be *4*. Also press the button to the **ON** state and set its **Mechanical Action** to **Latch When Pressed**. Save these settings by selecting **Make Current Values Default** in the **Operate** menu. Then save the VI under the name **Simple Read-Advanced VIs** in **YourName.llb.**

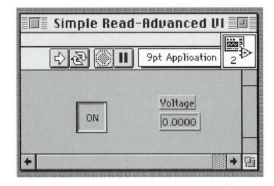

Voltage Acquisition Using the Advanced VIs

To gain an understanding of the details of an A/D operation, let's first construct a block diagram using the fundamental building-block VIs found in **Functions>>Data Acquisition>>Analog Input>>Advanced**. For a simple voltage reading, we will require only two Advanced Analog Input VIs—**AI Group Config.vi** and **AI SingleScan.vi**.

The Help Window for **AI Group Config.vi** is shown next. The job of this icon is to define the paths along which the analog-to-digital operations will be occurring. Thus the expansion-slot location of the DAQ board is established by a numerical input to **device** and the active analog input channels are listed in the form of an array of strings at the **channel scan list** input. This configurational information is needed by other Analog Input VIs to be subsequently called, so it is packaged together and output at **task ID**. Finally, all LabVIEW DAQ icons (including this one) have an error cluster input and output that you can elect to use for run-time error reporting (see the LabVIEW Online Reference or user manuals for more information). Because the error cluster is an input and an output terminal on all of the DAQ VIs, this parameter can be used to control the order of program execution through artificial data dependency.

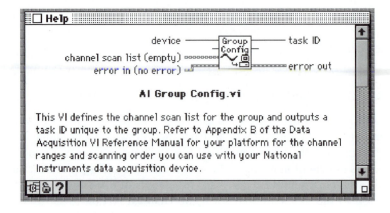

AI SingleScan.vi, whose Help Window follows here, performs the actual A/D data acquisition. Once presented with the necessary configurational information (including a list of active input channels) at its **task ID** input, this icon directs the DAQ board to digitize the signal at each of the active inputs channels. It then scales the acquired binary value(s) in Volts and outputs this data in an array at **voltage data**.

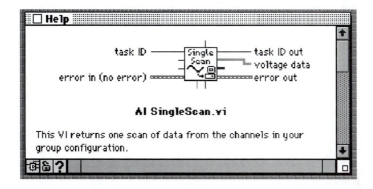

In this chapter, we will be monitoring only a single signal and so will only have one active channel. However, if in the future, you wish to digitize N signals through the use of N active channels it is important to understand that **AI SingleScan.vi** will not acquire the N digitized voltage levels simultaneously, but rather sequentially. This may disappoint you, but it follows from the fact that the DAQ board does not have N successive-approximation digitizers mounted on it, but only one. The multiple-channel capability is due to a multiplexer chip that properly sequences the various active channels (in the order that they appear in the **channel scan list** and at a rate determined by the *"channel clock"*) to the digitizer input. This sequencing operation results in an accurate voltage reading of all the N signals, taken over a succession of equally spaced times. In LabVIEWspeak, this sequencing operation, which results in one acquisition from each active channel input, is called a *scan*.

Code the following block diagram, which reads and displays the voltage at the **Channel 0** input every 0.5 seconds. Use **Build Array** to construct the required array of strings for **channel scan list** and **Index Array** to extract the index-zero array element from the **voltage data** array output. Program the integer **device** with the value you obtained for your DAQ board (for example, from **NI-DAQ Configuration Utility**).

Also, the correct manner of chaining together DAQ VIs for error reporting is shown. If an error does occur at one point in the chain, subsequent VIs will not execute and the error message will be passed to the **Simple Error Handler.vi**, which will display the message in a dialog box. **Simple Error Handler.vi** is found in **Functions>>Time & Dialog**.

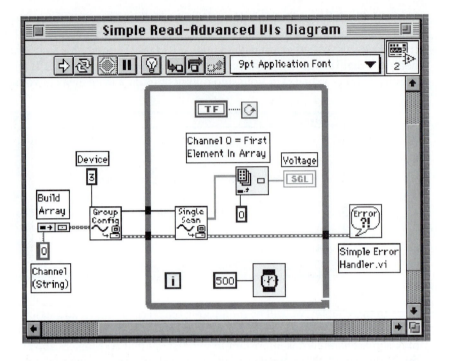

p.306)

Alternately, if you pop up on the **channel scan list** input and select **Create Constant**, an **Array Constant** formatted for strings will be presented to you. Simply enter the correct channel number as the index-zero element of this array.

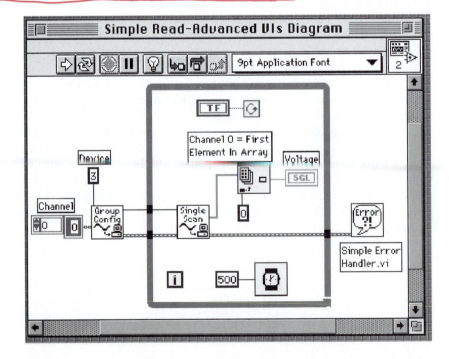

Connect a voltage difference of +5 Volts (or less) to the **Channel 0** analog input of your DAQ board. For example, on an E Series MIO-16 board in the differential mode, connect +5 Volts to pin 68 and GND to pin 34, while on a Lab-PC board in referenced single-ended mode, the **Channel 0** connection is between pin 1 and pin 9 (GND).

Run **Simple Read–Advanced VIs**. Does it read the input voltage correctly? Save your VI in **YourName.llb**.

Voltage Acquisition Using the Easy I/O VIs

An Analog Input Easy I/O VI called **AI Sample Channel.vi** performs essentially the same operation as the VI that you just wrote. A quick glance at its Help Window below indicates that it does so with extremely little effort required on your part.

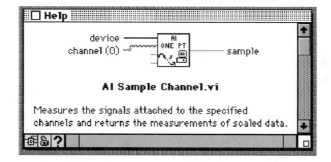

With **Simple Read–Advanced VIs** open, use **Save As…** to create a new VI called **Simple Read–Easy I/O**. Leave the front panel unchanged, but rewrite the block diagram using **AI Sample Channel.vi** from **Functions>>Data Acquisition>>Analog Input**.

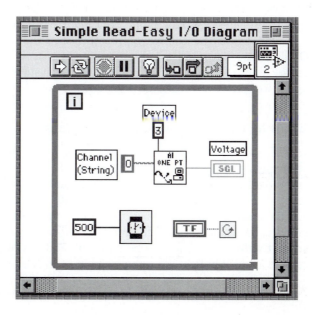

Verify that your new program accurately reads a voltage (within the range allowed by your DAQ board) at the **Channel 0** input.

Open **AI Sample Channel.vi** by double clicking on it in your block diagram. Look at its block diagram and open some of its subVIs. You will find that the Advanced Analog Input VIs are the basic building blocks from which **AI Sample Channel.vi** is constructed.

DIGITIZING OSCILLOSCOPE

In the previous exercise, you learned that each execution of **AI SingleScan.vi** acquires a single voltage reading of the analog signal at a channel input. By putting this icon within a repetitive loop and controlling the time per loop iteration with **Wait (ms)**, you wrote **Simple Read-Advanced VIs**, a program that, with the simple addition of some data storage and plotting capability, one might consider using to sample and display N data points of a time-varying waveform. The last sentence outlines the schematic for one of the most useful laboratory monitoring systems—the digitizing oscilloscope. Perhaps you've already thought of some cool things to do with this idea and are ready to go back and start to make the needed modifications to **Simple Read-Advanced VIs**... or perhaps you've already seen the Achilles' heel in this plan. Remember that **Wait (ms)** measures time by accessing a clock within your computer and, with an accuracy on the order of milliseconds, is pathetically imprecise. By using this icon to mark time in the data-taking process, the moment of each voltage digitization will be highly uncertain. With this large uncertainty in x-axis values, the envisioned oscilloscope's voltage versus time output might possibly work for very low-frequency inputs (say, less than 1 Hz), but will be useless over the higher range of frequencies that one would want an oscilloscope to operate.

Thankfully, a much better clock, with precision on the order of microseconds, exists on the National Instruments DAQ board plugged into your computer. And, even better, LabVIEW provides access and control of this clock for the purpose of data acquisition through several Advanced Analog Input VIs. Let's explore the function of these VIs as we write a useful digitizing oscilloscope program. The program we wish to write, called **Digital Oscilloscope**, will acquire and save N equally spaced voltage samples of a time-varying analog input, then quickly plot this array of data values. By repeating this process over and over, we'll achieve a real-time display of the waveform input.

Advanced VIs for a Buffered Analog Input Operation

In the program that we've set for ourselves, there is an initialization process needed at the outset. In this set-up routine, we must define for the computer the paths along which the analog-to-digital operations will be occurring and also set aside an appropriately sized block of memory for storage of the digitized data values. You already know how to use **AI Group Config.vi** to accomplish the former task, **AI Buffer Config.vi** (whose Help Window is shown next) can accomplish the latter. The **task ID** number, produced by **AI Group Config.vi**, must always be passed in and out of an Advanced DAQ VI. Then, to define a buffer in RAM that can store up to an N-element array of data values per active input channel, one need only wire a **Numeric Constant** of value N to **scans per buffer**. Remember: When using only a single channel input, one scan equals one voltage reading.

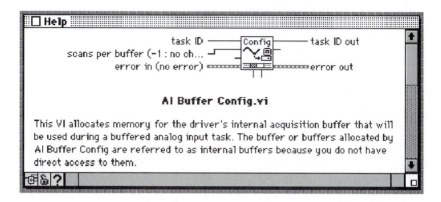

Next, we need three VIs that manage the actual data-acquisition processes on the DAQ board. The first of these three Advanced Analog Input VIs is **AI Clock Config.vi**, which allows you to set the frequency f_s at which data points will be discretely sampled. Simply wire the value of f_s (in units of Hertz) to its **clock frequency** input. You'll have to consult the manual for your particular DAQ board to find the maximum allowable value for this parameter. The AT-MIO-16E-10 model, for example, can digitize up to rates of about 100,000 Hz.

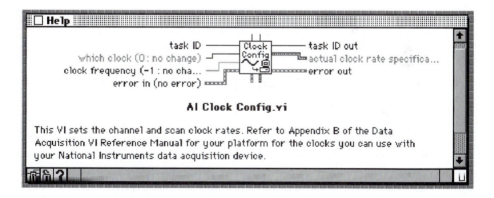

There are a couple of complications in configuring the data acquisition clock through the use of **AI Clock Config.vi**. First, some DAQ boards have two clock rates that can be set—the *scan clock* and the *channel clock*. When monitoring several AI channels "simultaneously," the scan clock controls the time between successive scans, while the channel clock determines the time delay between successive channel sampling within a given scan (Confusing, eh? Draw a picture; it's really not.). The programmer logs his or her choice of the clock-to-be-configured via the **which clock** input (1 = scan clock, 2 = channel clock). Many popular boards only have a channel clock, so I will define **which clock**

equal to *2* in the following program. You may have to alter this choice because of your particular DAQ board. Second, LabVIEW accommodates your requested clock frequency as best it can, but the DAQ board's clock cannot be run at just any arbitrary frequency. Your request is taken as a suggestion, and the closest available clock frequency is set on your board. You are supplied the result of this process in the output cluster **actual clock rate specification**. This cluster, when unbundled, provides the following parameters in this order—**actual clock frequency**, **actual clock period**, **timebase signal frequency**, **timebase divisor**. You will use the first and second of these parameters in your program. The third and fourth quantities describe how the clock frequency is actually produced on the board.

Next, **AI Control.vi** controls the analog input tasks and specifies the amount of data to be acquired during each cycle of operation. Wire the number of times N you wish the analog input signal to be digitized at the **total scans to acquire** input. This decision, along with the selected clock frequency, of course, determines the total time T over which you will observe the incoming waveform. The data acquisition operation that actually acquires these N data points is begun by providing the **control code** input with the value of *0* (Start). Once a data operation is under way, you can suspend it by wiring *1* (Pause Immediately) to the **control code** input. When you are through taking data, set **control code** to *4* (Clear). The icon will then stop the acquisition processes and free internal resources such as memory buffers.

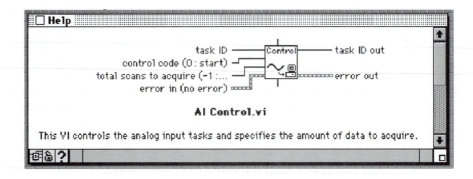

Finally, **AI Buffer Read.vi** retrieves the acquired data from the internal buffers and outputs it scaled in the units of Volts. Similar to the spreadsheet format, the output at **voltage data** is in the form of a 2D array with column-major order. That is, the equally spaced voltage readings acquired from a particular channel input reside in a single column, with the sequence of these data elements indexed by the row numbers. Sensibly, column-zero contains data from the first channel sampled in a scan, column-one contains data from the second channel, etc. In our case of a single active channel input, the 2D array output from **voltage data** is a single column and one may simply use **Index Array** to convert this single-column 2D array into a 1D array format (remember to **Disable Indexing** on the row index input). For future reference, if more than one channel is active, the resulting 2D array is transposed in comparison to the form in which LabVIEW's Waveform Graph accepts input data. This inconvenience can be easily remedied by activating **Transpose Array** in Waveform Graph's pop-up menu.

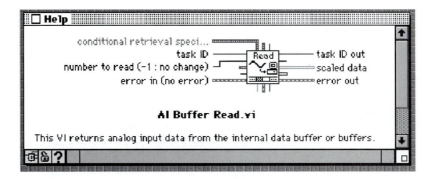

Analog and Software Triggering

As you know, when viewing a static-appearing image on an oscilloscope screen, you are not just looking at a single trace of data, but actually many traces that all lie on top of each other. If you are not careful to begin each trace at an equivalent point on the repetitive incoming waveform, the traces will not lie on top of each other, resulting in the undesirable situation of a moving image or, worse, a jumbled mess. The issue we are broaching here is called *triggering* and, for **Digital Oscilloscope** to be a useful program, we must build-in this capability.

In a commercially available oscilloscope, triggering is accomplished in the following manner. The input signal is monitored by an analog *"level-crossing"* circuit. The purpose of this circuit is to determine each time the incoming signal passes through a specified voltage level and then immediately trigger the scope's data acquisition process. By using knobs on the scope's front panel, you—the oscilloscope user—set the circuit's threshold level and specify whether acquisition should be initiated when the level is passed through starting from above (*negative*, or *falling, slope*) or starting from below (*positive*, or *rising, slope*) .

Now the bad news. Unless you have a specialized model, your National Instruments DAQ board probably doesn't have the analog level-crossing circuit described above. So somehow we must trigger **Digital Oscilloscope** by some other method.

Now the good news (but not as good as telling you that your DAQ board suddenly has analog triggering capability). **AI Buffer Read.vi** can operate in what is called the *conditional retrieval* mode. Here, the user specifies triggering conditions using a seven-element cluster input at **conditional retrieval specification**. When the icon retrieves the buffer of acquired data values into the computer's memory, it then searches (in software) from the beginning of the data array until it discovers the desired level crossing. Once the index of the array element satisfying the triggering condition has been determined, the preceding elements are snipped off and the correctly offset array is scaled and output at **voltage data**. There you have it— software triggering! Due to the finite digitizing rate of the DAQ board, software triggering has its limitations as you'll see, but it works well up to fairly respectable input frequencies.

Triggered Read VI

Construct the following front panel and save it under the name **Triggered Read** in **YourName.llb**. The control labeled **Slope** is an **Enumerated Type**, which can be found in **Controls>>List & Ring**. We learned a bit about this handy data-input format in the last chapter. Pop up on **Slope** to activate its **Digital Display** using the **Show** palette.

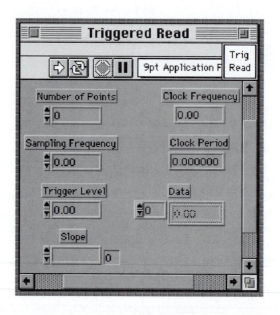

You now need to program **Slope** with the following sequence of three options: *no change*, *rising*, *falling*. This list of three (lowercase) choices is required by one of the Advanced Analog Input VIs—**AI Buffer Read.vi**—you will be placing on your block diagram. There are two methods for loading the required sequence of text messages into an **Enumerated Type** control. Here's a description of the easiest way. Using the Labeling Tool, type *no change* (using all lowercase letters) for the Enumerated Type's index-zero text message. Then press <*Shift-Enter*> (Windows) or <*Shift-Return*> (Macintosh) to store this element and also create the next (index-one) element. Type (lowercase) *rising* for the index-one text message, then press <*Shift-Enter*> (Windows) or <*Shift-Return*> (Macintosh) to enter this element as well as create the index-two element. Finally, type (lowercase) *falling* for the index-two text message, then (because no further elements need to be created), click on the **Enter** button in the toolbar. Alternately (and more tediously), after each text entry (that is, using **Enter** without the simultaneous pressing of the <*Shift*> key), you can pop up on the control and select **Add Item After** to create the next element. When you are through programming the **Enumerated Type** control **Slope**, use the Operating Tool to make sure it has only the following three entries: (0) *no change*, (1) *rising*, and (2) *falling*. If anything is wrong here, it will come back to haunt you later.

Assign the connector terminals consistent with the Help Window shown next, then save your work.

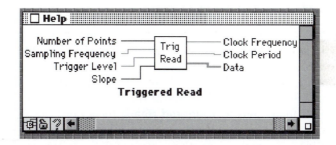

Using the Advanced Analog Input VIs described above, code the following block diagram. Wire an integer N to the **scans per buffer** input of **AI Buffer Config.vi**, where N is greater (for software triggering, it's wise to be 2.5 times greater) than the number of elements in the largest data array you expect to acquire. In the **conditional retrieval specification** cluster, **Trigger Channel (I32)** specifies which channel input to search to find the trigger condition, **Slope** tells whether to search for a rising (=1) or falling (=2) voltage level, **Trigger Level (SGL)** specifies the trigger threshold in Volts, **hysteresis (SGL)** is a noise suppression filter familiar to those who have worked with Schmitt Trigger circuits, **skip count (I32)** is the number of specified conditions to skip before "triggering" the acquisition, and **offset (I32)** indicates the position the VI begins reporting data relative to where the triggering condition was found (allowing for pre- and post-triggering). If you get a bad wire when connecting the bundle to **AI Buffer Read.vi**, there's probably something wrong with the text in the **Slope** control (e.g., an extra <*Space*> character at the end of a string).

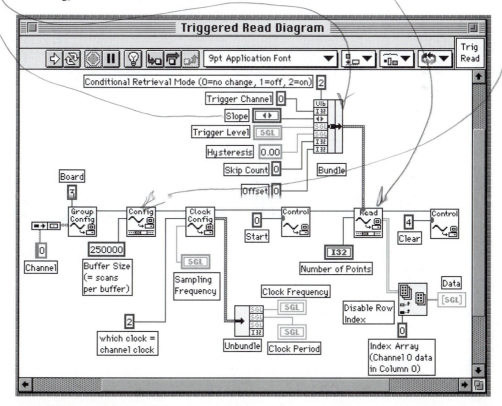

The Enumerated and Ring Constants

In the previous diagram, I "manually wired" a **Numeric Constant** (with Representation of **U16**) to the **which clock** and **control code** inputs of **AI Clock Config.vi** and **AI Control.vi**, respectively. As you can see, gray coercion dots appear at both of the **control code** inputs, indicating that LabVIEW is being forced to accommodate for data-type incompatibilities. Opening **AI Control.vi** as well as **AI Clock Config.vi** (e.g., by double clicking on their icons in **Triggered Read**) for a quick inspection of their block diagrams, one finds that the **control code** input is an **Enumerated Type**, while **which clock** is a **Text Ring** (a control with much in common to the Enumerated Type). By deleting the manually wired **Numeric Constants** and, instead, popping up on these inputs and selecting **Create Constant**, the appropriate **Enumerated Constants** (for **control code**) and **Ring Constant** (for **which clock**) will be automatically wired, as shown here.

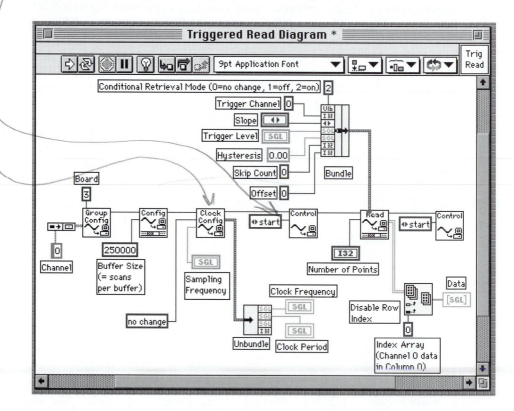

When created in this manner, the **Ring Constant** and **Enumerated Constants** display the default value for the input to which they are wired. However, a selected value can be easily changed. As an example, place the Operating Tool over the **Ring Constant** wired to **AI Clock Config.vi**.

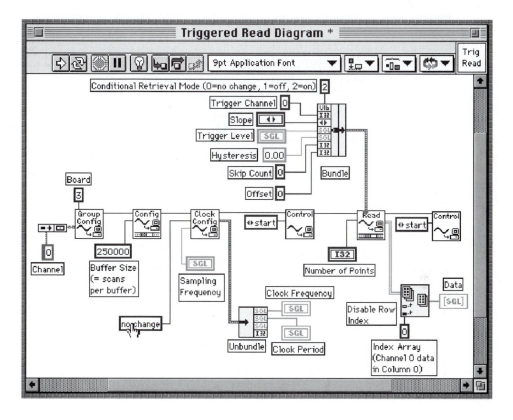

Click and select the desired value from the list of items that appears.

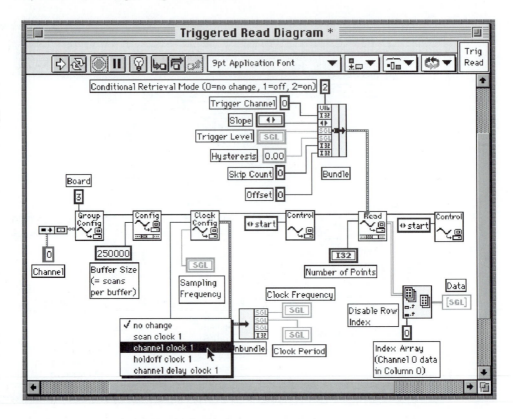

Use this method to select the correct clock for your system (in most cases, the channel clock) and to instruct **AI Control.vi** to *Start* and *Clear* at the appropriate locations on your diagram.

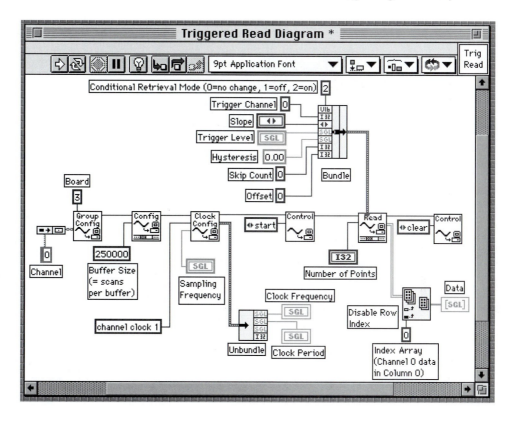

If you'd like, try removing the coercion dot at the **conditional retrieval specification** input of **AI Buffer Read.vi.** The data-types expected in the **conditional retrieval specification** cluster (which can be discovered by looking at the block diagram for **AI Buffer Read.vi**) are shown here.

The **Enumerated Constant** required for the **Bundle**'s first input is found in **Functions>>Numeric**. Once placed on the block diagram, you must program it (in exactly the same manner that you programmed the front panel **Enumerated Type** control) with the following three lowercase elements: *no change, off, on*. Once programmed, select *on*. The final diagram will appear as follows.

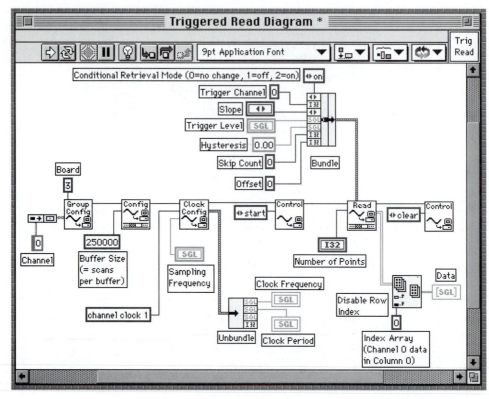

Remember, it is not imperative that you perform these described mouse gymnastics to remove coercion dots. LabVIEW (don't take this personally) is idiot-proof in the sense that it will automatically make the proper format conversions for you, correcting the data-type mismatches that you inadvertently hardwire into your VI. However, data-type conversions do take time and this needless loss of time should be of special concern in a data acquisition program. Thus, it's worth some effort on your part to eliminate as many of those pesky gray dots from your block diagrams as possible.

Number of Samples VI   p. 304

To use **Triggered Read**, you will need to be able to tell it how many data points to acquire. Write the following VI called **Number of Samples** to make this calculation and save it in **YourName.llb**. For our oscilloscope, we will offer the user the choice of four total times T over which to digitize the incoming waveform: *1 msec, 10 msec, 100 msec,* and *1 sec*. Put an **Enumerated Type** on the front panel and enter its item-0 through item-3 as these four possibilities.

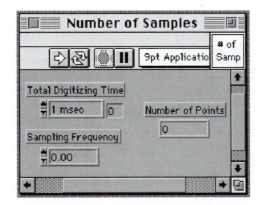

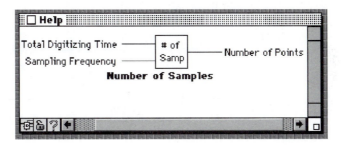

Code the block diagram as shown below. Here the sampling period Δt is given by

$$\Delta t(\text{in msec}) = \frac{1000}{f_s}$$

and so to digitize a waveform for a total time T (in msec), the **Number of Points** N is

$$N = \left[\frac{T}{\Delta t} + 1 \right]$$

where, for our VI, T can be 1 msec, 10 msec, 100 msec, or 1000 msec (=1 sec). Note another convenient feature of the **Enumerated Type**: When wired to a Case Structure's **selector terminal**, the various Case windows are identified by the **Enumerated Type**'s descriptive text strings, rather than by the associated integers (whose meanings are less transparent).

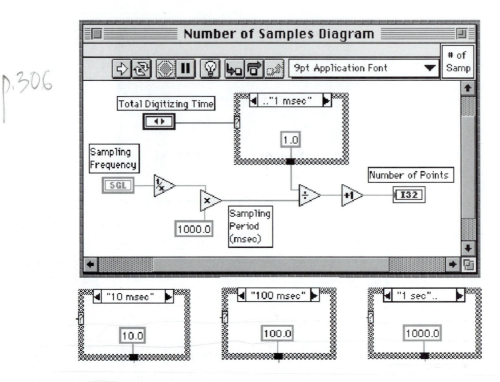

p.306

Save **Number of Samples**.

Digital Oscilloscope VI

You are now ready to write the digitizing scope program. Open a new VI, switch to the block diagram, and place **Number of Samples** and **Triggered Read** on it using **Functions>>Select a VI**... One at a time, pop up on the four inputs shown and select **Create Control**.

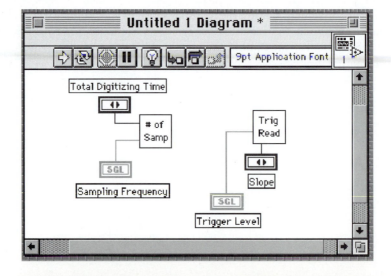

Return to the front panel, arrange the four controls that appear there in some pleasing pattern, then complete the panel as shown. Using the Waveform Graph's Palette, increase the precision of x-axis labeling to *4* so that good labeling will be obtained even when the total digitizing time is 1 millisecond. If you wish, customize the values that appear in the front-panel controls, then save them by selecting **Make Current Values Default** in the **Operate** menu. Save the VI under the name **Digital Oscilloscope** in **YourName.llb**.

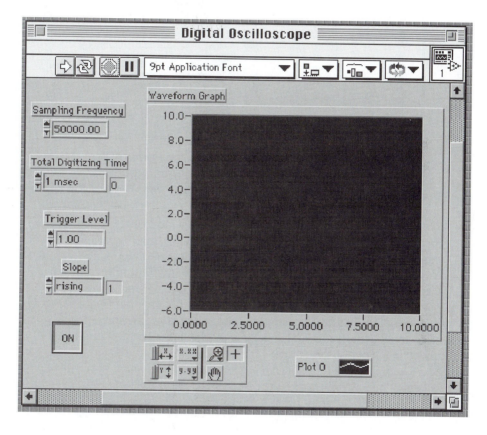

Finally, complete the block diagram as illustrated next

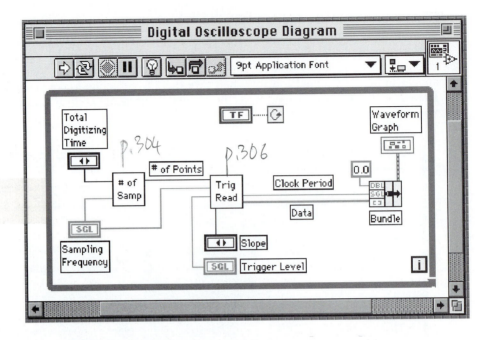

p.289,
290

Let's explore the operation of **Digital Oscilloscope**. Using a function generator, input a sinusoidal signal at the **Channel 0** Analog Input and view it on your VI's front panel. Set **Sampling Frequency** at the maximum allowed value for your DAQ board. You should find that low-frequency signals are faithfully reproduced by **Digital Oscilloscope**, but that distortion arises as the signal frequency is increased (due to the paucity of sampled points per sine-wave cycle at high input frequencies). For input frequencies above half of the sampling rate, look for the aliasing effect.

Input a low frequency signal and play with the **Trigger Level**. Make sure that you understand your observations.

Now, increase the signal frequency and you will find that the displayed waveform becomes "jittery." This observation can be explained as follows. For an input sinusoid of frequency f and amplitude A described by $x = A \sin(2\pi ft)$, its change during one sampling period Δt is given by

$$\Delta x = \frac{dx}{dt}\Delta t = 2\pi fA\Delta t \cos(2\pi ft) \tag{4}$$

Calculate the magnitude of this change

$$|\Delta x| = 2\pi fA\Delta t \tag{5}$$

in the frequency range where the jittering is observed. From your knowledge of software triggering, can you now explain the jitter? Can you suggest what value of **Trigger Level** will minimize jittering at high frequencies? Test your prediction.

If you encounter triggering problems at lower frequencies (due to noise in the input signal), try increasing the **hysteresis** value (given in Volts) within **Triggered Read** from *0.00* to something larger like *0.10*. Equations (4) and (5) can also be used to explain why this works.

Improving the Triggered Read VI

Finally, you may have noticed a gross inefficiency in our **Digital Oscilloscope** algorithm. Namely, several subVIs contained within **Triggered Read** simply set up (**AI Group Config.vi**, **AI Buffer Config.vi**, **AI Clock Config.vi**) or shut down (**AI Control.vi** with *Clear* selected) data acquisition operations and, thus, need only be called during one, or a small subset, of the multitude of **Digital Oscilloscope's** While Loop iterations. However, as now written, these set-up and shut-down (collectively termed *"overhead"*) operations are called each time **Triggered Read** executes. Since your data-taking system is not able to digitize the incoming stream of real-time data during overhead operations, needlessly including them increases the program's *dead time*, the percent of each iteration period during which the system is not sensitive to the incoming data. With increasing dead time, more and more cycles of the incoming repetitive data will stream past your system's input undetected. In certain situations, such as when collecting a large number of data-cycles with the intent of adding them together so as to average-out random noise, a programming inefficiency that causes you to miss a significant percentage of the incoming data-cycles can easily extend the completion time of your experiment by hours (or sometimes even days).

If you care to correct the above-described problem, modify the front panel of **Triggered Read** to include two new inputs—a **Digital Control** (**I32**) labeled **Iteration** and a **Rectangular Stop Button**. Then store this front panel as a new VI named **Smart Read** by using **Save As**....

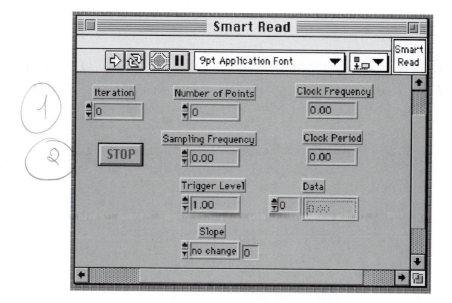

Assign the connector consistent with the following Help Window.

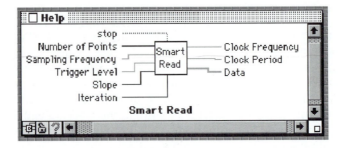

Then rewrite the block diagram so that **AI Group Config.vi** and **AI Buffer Config.vi** are called only when **Iteration** equals *0*, and **AI Clock Config.vi** only when the value of **Sampling Frequency** is changed. Since a change to **Sampling Frequency** can occur when the system is busy taking data, you must call **AI Control.vi** with *Pause Immediately* selected, prior to requesting **AI Clock Config.vi** to change to the clock frequency. Finally, we want **AI Control.vi** with *Clear* selected to be called only during the last data-taking iteration when the Boolean **Stop Button** is FALSE.

The following block diagram accomplishes the desired features. The While Loop is placed on this diagram solely for the use of its shift registers, which remember the values of important numerics from one execution of **Smart Read** to the next.

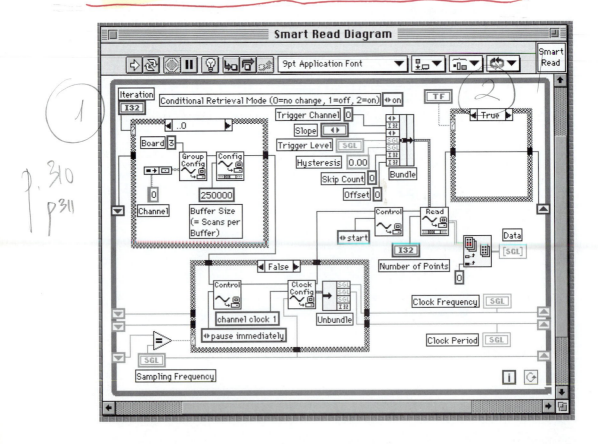

The other Case windows are as follows. The **Default** window for the numeric **Case Structure** is defined by popping up on the window's control.

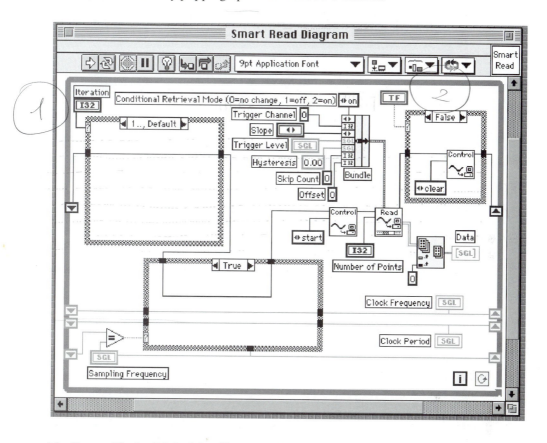

Finally, modify the **Digital Oscilloscope** diagram as shown below. Then run this VI and see if it works.

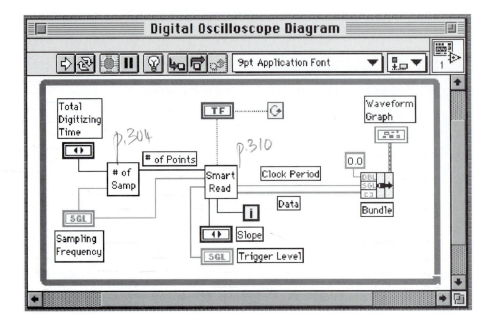

Save the final versions of your VIs as you close them.

SPECTRUM ANALYZER

Open **FFT–Magnitude Only** in **YourName.llb**, and using **Save As**… create a new VI called **Spectrum Analyzer**. Add an ON/OFF button to the front panel.

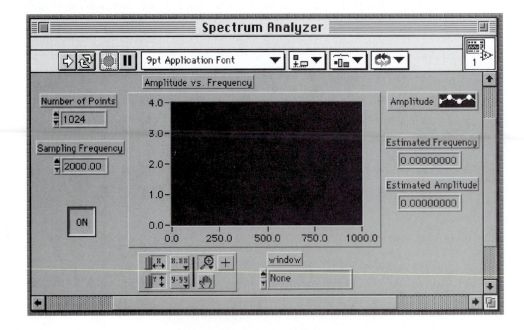

Switch to the block diagram. Replace **Sine Wave** with **Triggered Read** (or **Smart Read**), then complete the modifications shown.

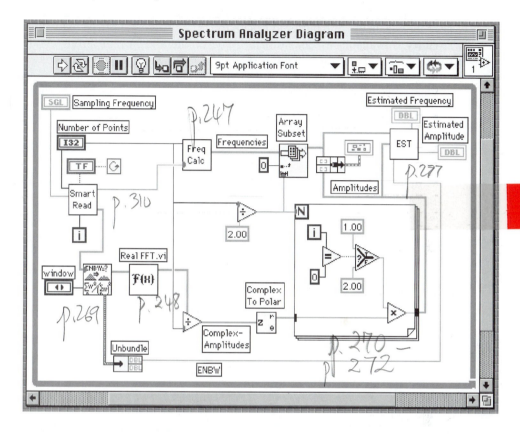

Input a sine wave from a function generator into the **Channel 0** input of your system. With **Number of Points** and **Sampling Frequency** set to *1024* and *50000*, respectively, run **Spectrum Analyzer** to see the spectral make-up of your signal. How do the values of **Estimated Amplitude** and **Estimated Frequency** compare with the known values of your input signal?

Increase the input sine wave's frequency beyond the Nyquist frequency limit to see the aliasing effect. Explain your observations using Equation (3).

Try inputting a square wave. Fourier analysis of a unit-amplitude square wave of frequency f predicts

$$\text{Square Wave} = \frac{4}{\pi} \sum_{n\,\text{odd}} \frac{1}{n} \sin(2\pi n f t)$$

Is the output of **Spectrum Analyzer** consistent with this result? Do you note any aliasing of the square wave's higher harmonics?

DIGITAL THERMOMETER

Build the following circuit, choosing resistance values so that a constant current of 0.1 mA flows through the thermistor. You may wish to include the high input-impedance unity-gain op-amp buffer to isolate the thermistor circuit from the voltage-sensing circuitry. If available, use an ammeter to determine the current through your thermistor precisely. It probably will differ slightly from 0.1 mA due to resistor tolerances.

Connect the circuit's V_{out} and GND to your LabVIEW system's I/O Connector so that you are able to monitor the voltage difference across the thermistor. Then, write a LabVIEW program called **Digital Thermometer**, which executes the following sequence of actions:

- perform an A/D operation that reads the thermistor's voltage difference into the computer.
- determine the thermistor's resistance via Ohm's Law using the known thermistor current.
- calculate the thermistor's temperature via the Steinhart–Hart Equation with the coefficients solved for in Chapter 8.
- display the calculated temperature in °C.
- wait a predetermined amount of time (say, 0.5 or 1 second), then repeats the above sequence.

Use **Digital Thermometer** to measure the room's temperature and your body's temperature.

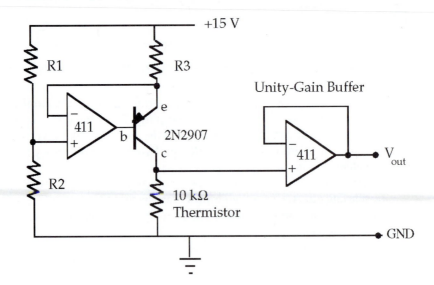

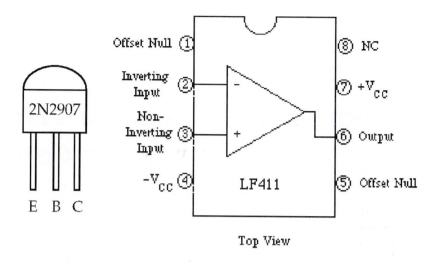

Top View

CHAPTER 11

Digital-to-Analog Conversion and PID Temperature Control

In this chapter, you will learn how to perform a digital-to-analog (D/A) operation using the National Instruments DAQ board within your computer. With this skill in place, you will then apply your LabVIEW programming skill in the construction of a feedback-based system that controls the temperature of an aluminum block with great precision.

Digital-to-Analog Circuitry

How is a D/A operation actually accomplished by the DAQ board's hardware? Well, in schematic form, a D/A converter is an inverting op-amp amplifier with a gain that can be programmed digitally. The inverting amplifier shown here is well known to students of basic op-amp circuitry.

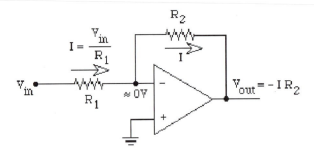

By applying the two "golden rules" that accurately describe op-amp behavior in such a feedback circuit—the voltage difference between the two inputs is zero and the inputs draw no current—it is easy to prove that the inverting amplifier's output voltage is given by

$$V_{out} = -V_{in}\frac{R_2}{R_1}$$

Thus, if V_{in} and R_2 are fixed, V_{out} can be varied by varying R_1. This is exactly the situation in a D/A chip, where V_{in} is a reference voltage V_{ref} and R_1 is a digitally-controlled resistance whose value ranges in discrete steps from a maximum of ∞ down to a minimum of just-slightly-greater-than R_2. The chip then is able to produce a finite number of digitally selectable voltages in the range of 0 V to (almost) $-V_{ref}$. This range can be made bipolar with an additional DC offset circuit, if desired.

In a D/A chip, the above-mentioned resistance R_1 is actually many resistors configured in what is known as an *R/2R ladder network*. The next figure shows how such a network would appear in a 4-bit D/A chip. Here, the tail of each resistor of value 2R is attached to an analog switch and each of these switches is controlled by the value of a TTL level (not shown). By setting a particular TTL control signal in its HIGH or LOW state, one can then funnel the current through the associated resistance 2R either to the negative op-amp input or ground, respectively. Note that the op-amp's inverting input is at virtual ground (i.e., only a tiny fraction of a millivolt above 0 Volts), so the flow of current through a particular resistor 2R is unchanged when its associated analog switch toggles from one of its states to the other.

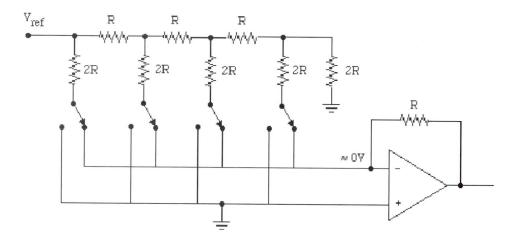

The symmetry of the R/2R ladder allows us to determine the current through each of its resistors in the following manner. Starting with the two rightmost resistors, we note that each resistor's tail is connected to ground (independent of the state of the rightmost analog switch). The other ends of these resistors are connected at node **A**, so the incoming current at this node must, by symmetry, be equally split between the two. Call the current through each resistor I′.

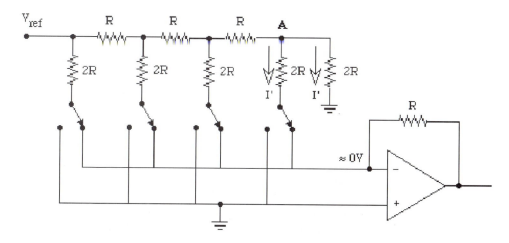

The incoming current then at node **A** is $2I'$. Now note that the previously-discussed pair of 2R resistors is a parallel combination with an equivalent resistance of R. This equivalent resistance is in series with the horizontal resistor R connected at node **A**, so the boxed resistance shown below is equivalent to a resistance of 2R with its tail at ground and a total current of $2I'$ flowing through it.

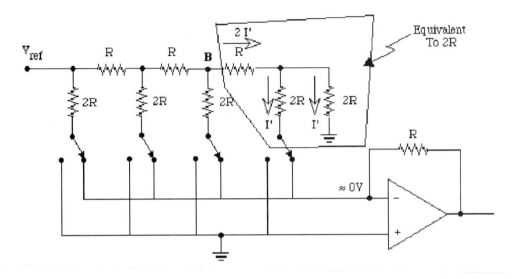

The equivalent resistance 2R with current $2I'$ (highlighted) is in parallel with the single 2R resistor connected at node **B**. By symmetry, a current of $2I'$ must flow through the single 2R resistor. At node **B**, then, we have two 2R resistances in parallel connected in series with a horizontal resistor R. This entire combination has an equivalent resistance of 2R. This equivalent resistance, shown in the following box, has its tail at ground and a total current of $4I'$ flowing through it.

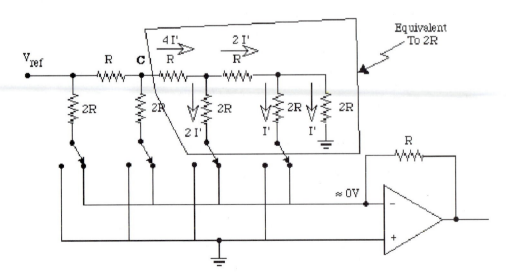

The above-highlighted equivalent resistance of 2R is in parallel with the single 2R resistor connected at node **C**. By symmetry 4I′ flows through this single 2R resistor, so a total current flows in the horizontal resistor R attached at **C**. The resistors boxed below then are equivalent to a resistor 2R with its tail at ground and a total current of 8I′ flowing though it. From symmetry, a current of 8I′ flows through the single resistor 2R connected at **D**, so a total current of 16I′ flows out of the reference voltage source.

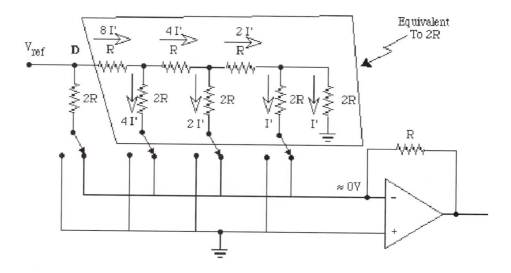

In terms of I′, the current we have determined in each resistor of the ladder is given here.

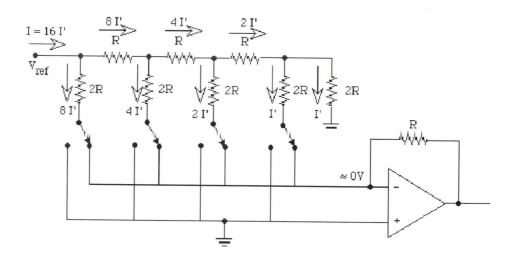

Noting that the reference voltage source "sees" an equivalent resistance of R, the total current I equals V_{ref}/R. Rewriting the currents in terms of I, we discover the following flow of currents though the ladder.

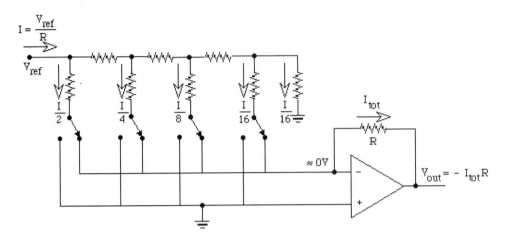

Thus, if all of the analog switches are selected to be in their LOW state, no current will flow toward the inverting input of the op-amps and the output voltage will be $V_{out}^{min} = 0$. Conversely, if all of the analog switches are selected to be in their HIGH state, the maximum current I_{max} will be diverted to the op-amps's inverting input. From the diagram,

$$I_{max} = \left(\frac{1}{2} + \frac{1}{4} + \frac{1}{8} + \frac{1}{16}\right)I = \frac{15}{16}I = \frac{15}{16}\frac{V_{ref}}{R}$$

so the maximum output voltage is

$$V_{out}^{max} = -I_{max}R = -\frac{15}{16}V_{ref}$$

The 14 other permutation of states for the analog switches will produce a succession of 14 other equally spaced output voltages that are bracketed by these extremal values.

By generalizing this argument, it is easy to prove that an n-bit D/A device contains an R/2R ladder with $(n + 1)$ 2R-resistor rungs and produces 2^n equally spaced output voltages in the range from 0 to $-(1 - 2^{-n})V_{ref}$. Commercially available devices typically have n = 8, 10, 12, 14, or 16. For the typical value of $V_{ref} = 10$ Volts, a 12-bit D/A device's *resolution* (i.e., spacing between successive output voltages) is $\Delta V = V_{ref}/2^n = 2.44$ mV.

SIMPLE DIGITAL-TO-ANALOG OPERATION

Build a VI that instructs your DAQ board to perform a D/A operation. For this simple task, the Easy I/O DAQ VI called **AO Update Channel.vi** will do the job. This icon is found in **Functions>>Data Acquisition>>Analog Output**. Its Help Window is shown next.

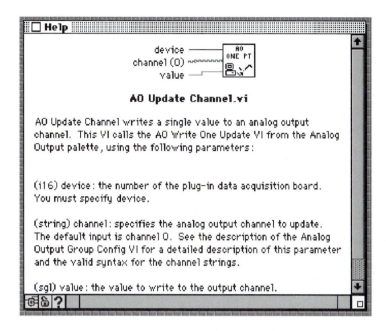

The D/A VI will appear as follows. Name it **Simple Write–Easy I/O** and save it in **YourName.llb**. You may wish to design an icon and assign the connector, as shown, so that this program can be used as a subVI in the temperature control VI to be written shortly.

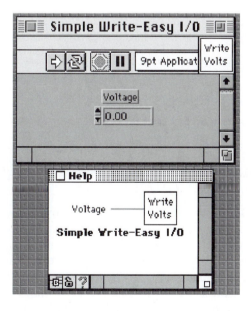

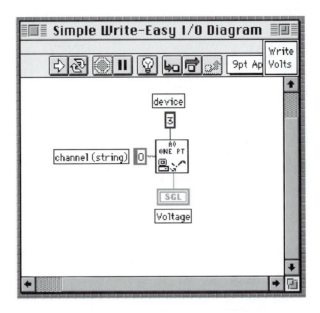

Attach a voltmeter between your DAQ board's Digital-to-Analog **Channel 0** Output (DAC0OUT) and Analog Output Ground (AOGND), then verify that you can use **Simple Write–Easy I/O** to output voltages in the range appropriate to your board. The values of these board-specific parameters are listed for some popular DAQ boards in the following table.

Board	D/A Voltage Range (Maximum)*	DAC0OUT (pin)	DAC1OUT (pin)	AOGND (pin)
E Series MIO-16E	± 10 Volts	22	21	54, 55
PCI-1200, Lab-PC	± 5 Volts	10	12	11

*Other ranges are software-selectable.

VOLTAGE-CONTROLLED BI-DIRECTIONAL CURRENT DRIVER FOR THERMOELECTRIC DEVICE

Build the following circuit. It is designed to provide the rather large bi-directional current flow necessary to power both the heating and cooling capabilities of the thermoelectric (TE) device. The voltage level V_{in} serves as a "selector code word," which dictates the magnitude and direction of the current flow through the TE module. If the power source that supplies the ±8 Volts in your circuit has built-in ammeters, use them to monitor the current flow through the TE device, eliminating the need to insert the ammeter shown in the diagram. See Appendix I for a discussion of important practical issues regarding this circuit, especially the need (and method) for dissipating the prodigious amount of heat it produces.

Use **Simple Write–Easy I/O** to apply various voltages at V_{in}. First, test that when V_{in} is positive (negative), an electric current flows in the proper direction through the TE device to heat (cool) the aluminum block in your experimental set-up. If a positive V_{in} cools the block, reverse the connections to the TE device within your circuit. Second, find the positive and negative voltages V_{sat}, where V_{sat} is the minimum voltage at V_{in} for which the current to the TE device becomes saturated. The value of this saturation current is determined by the capability of the power supplies attached to the TIP transistors. We then define the range of acceptable values for V_{in} to be between $\pm V_{sat}$.

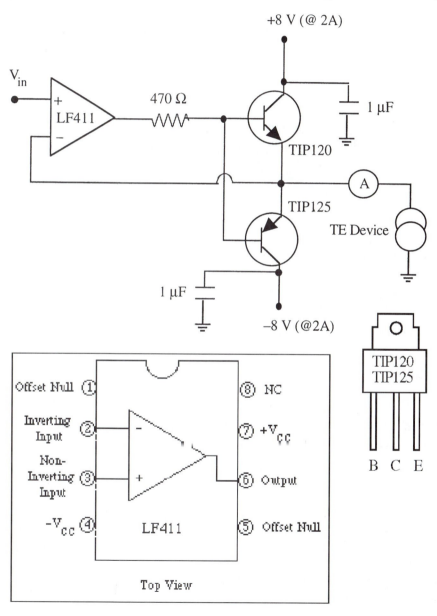

PID TEMPERATURE CONTROLLER

Build a digital temperature controller that can control the temperature of an aluminum block to within less than 0.05°C of a given set-point temperature. The set-point temperature can be chosen anywhere within the range of 0 to 40°C.

An algorithm for such a controller is as follows: (1) read the block's temperature T_{sample} using a thermistor as the temperature sensor, (2) compare T_{sample} with the desired set-point temperature $T_{set-point}$, (3) based on this comparison, decide what value of V_{in} will most appropriately command the TE device to provide the heating or cooling needed to bring T_{sample} closer to $T_{set-point}$, (4) apply this voltage at the V_{in} input of the voltage-controlled current driver circuit shown above, (5) repeat this process continuously to obtain the desired temperature control of the aluminum block.

Control Methods

The *Proportional Control* method offers a simple procedure for deciding what voltage should be applied at V_{in} in your temperature-control algorithm. In this method, the difference between the desired set-point temperature $T_{set-point}$ and the actual sample temperature T_{sample} is defined to be the error $E \equiv T_{set-point} - T_{sample}$. Then the control voltage V_{in} is simply taken to be directly proportional to E:

$$V_{in} = A\ E\ \text{(Proportional Control)} \qquad (1)$$

where A is a constant called the *gain*. The value for the gain is chosen empirically such that, when $T_{set-point}$ is changed, the optimal value for A will cause the system to ramp to the new set-point and then stabilize near there (see the following discussion) quickly .

Here is a practical consideration you'll confront when trying to implement Equation (1): When the sample temperature is far from the set-point, the error E may become large enough to generate a calculated value for V_{in} that falls outside its acceptable range of $\pm V_{sat}$. The solution to this problem is to truncate the expression given in Equation (1) so that the magnitude of V_{in} is never allowed to be greater than V_{sat}. The graphical representation for the Proportional Control algorithm is as follows.

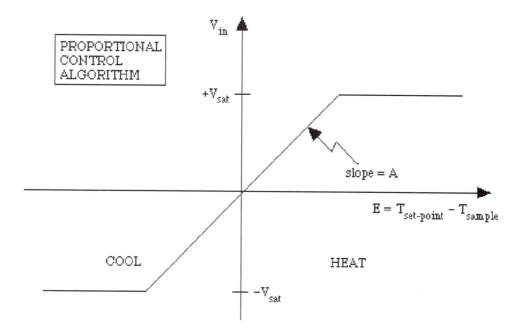

Although pleasingly simple, the Proportional Control method is intrinsically flawed. Here's why. Assume the sample is initially at room temperature and that you select $T_{set-point}$ to be above room temperature. Turning on the Proportional Control algorithm, the initial error E will be positive, resulting in a command to the TE device to heat the sample. So far so good. As time goes on, the algorithm will issue the proper heating instructions to bring the sample closer and closer to the desired set-point temperature. When T_{sample} nears $T_{set-point}$, the positive error E will become small, causing the Proportional Control to ease off on the applied heating power so that $T_{set-point}$ is gently approached. Then, at the decisive moment when T_{sample} equals $T_{set-point}$, the error E becomes zero and the Proportional Control turns off power to the TE device. There is the flaw. Unfortunately, when at an elevated temperature, the sample will constantly be losing heat to its surrounding (room-temperature) environment through the heat-transferring processes of conduction, convection, and radiation. Thus, to maintain a sample at a set-point above room temperature, heat must constantly flow into the sample to counteract the heat it is losing to its environment. Since the Proportional Control turns off the heating when T_{sample} equals $T_{set-point}$, the sample will never be able to stabilize at the desire temperature. Rather, the sample will stabilize at some temperature $T_0 < T_{set-point}$. The positive error E produced at the magic temperature T_0 commands just the right amount of heating by the TE device to cancel out the heat being lost by the sample to the environment at that temperature. By similar reasoning, a set-point below room temperature will result in the sample stabilizing at a temperature $T_0 > T_{set-point}$.

Luckily, there is an easy remedy to the above-described defect in the Proportional Control algorithm—simply include a constant term V_0 on the right-hand-side of Equation (1):

$$V_{in} = A E + V_0 \tag{2}$$

When this expression is implemented, the proportional term will ramp the sample to the set-point temperature as before. Then, once at the set-point (where $E = 0$), V_0 will instruct the TE module to provide the constant heating (or cooling) necessary to counteract heat losses (or gains) to the environment, thereby stabilizing the sample at $T_{set-point}$. Of course, the value of V_0 must be precisely chosen so that the environmental influences on the sample, when at $T_{set-point}$, are perfectly neutralized.

Now the best news: There is an elegant way to build intelligence into the control algorithm so that the correct value for V_0, appropriate to the chosen set-point, will be found automatically. In the *Proportional-Integral (PI) Control* method, rather than defining V_0 as a fixed constant, an integral term is used to construct the correct constant during run time according to the following expression:

$$V_{in} = A E + B \int E \, dt \quad \text{(PI Control)} \tag{3}$$

Here, the integral keeps a running sum of all the error values that occur over the entire execution time of the algorithm. During the times when T_{sample} is below $T_{set-point}$, a positive contribution will be made to the sum. When T_{sample} is above $T_{set-point}$, a negative contribution will be made. Due to the self-correcting manner in which these contributions are made, the second term in Equation (3) will eventually converge to the constant V_0, which allows the sample to stabilize at the set-point $T_{set-point}$. From that point on, the error E will equal zero and the value of the integral (and thus constant V_0) will no longer change.

Inclusion of a derivative term to damp out oscillations provides one further refinement to the control algorithm. The expression for this so-called *Proportional-Integral-Derivative (PID) Control* algorithm then is given by

$$V_{in} = A E + B \int E \, dt + C \frac{dE}{dt} \quad \text{(PID Control)} \tag{4}$$

where A, B, and C are constants. For the experimental situation of discrete data sampling, where the error E is determined every Δt seconds, the value of the control voltage V_{in} after the nth sampling can be approximated by

$$V_{in} = A E + B \Delta t \sum_{m=0}^{n} E_m + \frac{C}{\Delta t} \{E_n - E_{n-1}\} \quad \text{(Discrete PID)} \tag{5}$$

where the summation in the second RHS term is over *all* the error-values determined since the control algorithm was turned on.

Temperature Control System

To construct your digital temperature-control system, start by rebuilding the thermistor-based digital thermometer from Chapter 10. The thermistor, which is imbedded in the aluminum block, will be used to monitor T_{sample}. If needed, add the 1 μF capacitor as shown below to shunt high-frequency noise signals to ground and the unity-gain buffer to prevent the voltage-sensing circuitry from loading the thermistor circuit at low temperatures (when the thermistor resistance is large). Then connect V_{out} and GND to the inputs of an A/D channel of your LabVIEW system.

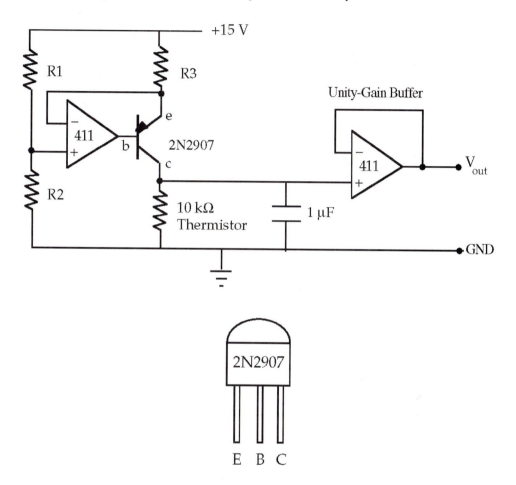

Now code a temperature control VI called **Temperature Controller** according to the following guidelines.

A. Sample-Temperature Reading: Noise-free measurements of the sample's temperature will be a "must" in order for your temperature-control algorithm to function properly. To average-out random fluctuations when reading the thermistor voltage, you might try the following: At each sampling of the temperature, read the thermistor voltage a number of times, then analyze this data to determine the mean (i.e., DC level). The Easy I/O Analog Input VIs, found in **Functions>>Data Acquisition>>Analog Input**, make acquiring the array of voltage measurements simple: either put **AI Sample Channel.vi** in a For Loop or else read an array of voltage levels into a buffer using **AI Acquire Waveform.vi**. To do the job of finding the DC level from this data, look at the VIs available in the subpalettes of **Functions>>Analysis**. Any one of several VIs there will perform the required calculation. Once you have a reliable value for the thermistor voltage, use Ohm's Law and the Steinhart–Hart Equation to convert it to the temperature T_{sample}.

   B. Control Algorithm: To decide the proper heating or cooling needed by the TE device, implement the Discrete Sampling PID Control algorithm in software. Remember that Δt in Equation (5) is the time between successive determination of the error E; think carefully about what this time is in your particular VI. Also make sure to include a truncation feature, that is, if Equation (5) yields a value of V_{in} with magnitude greater than V_{sat}, then truncate the magnitude of V_{in} to V_{sat}. You will have to determine the optimum value for the constants A, B, and C empirically. If your set-up is similar to the one described in Appendix I, try $A = 7, B = 1, C = 0.5$.

   C. Output to the TE Device: Use **Simple Write-Easy I/O** as a subVI on your block diagram to output the PID-determined value for V_{in} over one of LabVIEW's D/A channels to the input of the TE device's voltage-controlled current driver circuit.

   D. Front Panel: Use your creativity here. Remember you have a wide array of controls, indicators and graphs available that can be used, for example, to monitor E and T_{sample}.
   When you get the temperature controller to work, show it off to your friends and instructor!

CHAPTER 12

GPIB–Control Of Instruments

In previous chapters, you have used LabVIEW software to transform a personal computer (equipped with an appropriate National Instruments DAQ plug-in board) into several handy laboratory instruments. In particular, you programmed this system to become a voltmeter, digitizing oscilloscope, spectrum analyzer, digital thermometer, voltage source, and temperature controller. Pause to consider the following tantalizing prospect: Perhaps the only instrument required in a modern-day laboratory is a DAQ board-equipped computer controlled by LabVIEW software. That is, by simply writing a collection of appropriate VIs, it might be possible for you—the contemporary scientific researcher—to satisfy all of your laboratory instrumentation needs with this single LabVIEW-based data acquisition system. This system's tremendous flexibility would then obviate the need to purchase an expensive collection of stand-alone electronic equipment such as power supplies, function generators, picoammeters, spectrum analyzers, and oscilloscopes.

The functioning VIs that you have written in Chapters 10 and 11 demonstrate that this "tantalizing prospect" can, at least in certain situations, be realized. But don't discard your stand-alone instruments just yet. The timing, speed, sensitivity, and simultaneous data-taking requirements of many contemporary research experiments are beyond the capabilities of your DAQ board. For instance, while the LabVIEW-based digitizing oscilloscope we constructed worked well for observing audio-range frequencies (less than 20 kHz), it would prove miserably inadequate at displaying the several nanosecond-wide voltage pulses emanating from a photomultiplier tube. In this latter situation, a stand-alone digitizing scope with a very fast analog-to-digital converter (on the time scale of 1 gigasamples per second) would do the job nicely. Thus stand-alone instruments play a central role in state-of-the-art research, and so it might not surprise you to find that they, too, fall within the scope of LabVIEW.

The GPIB Write and GPIB Read Operations

Over the past few decades, a communications standard called the *General Purpose Interface Bus (GPIB)* has evolved by which stand-alone instruments can be software-controlled using a personal computer. In this communications scheme, a particular instrument obeys an array of manufacturer-defined ASCII character commands that represent all the possible ways of manually pressing buttons, turning dials, and viewing output data on its front panel. Through *GPIB Write* operations, these ASCII commands can be sent from a computer to the instrument to configure it properly for a data-taking measurement. After the measurement is complete, the resultant data can then be retrieved from the instrument into the computer using a *GPIB Read* action.

LabVIEW facilitates GPIB Write and GPIB Read operations with two built-in VIs—
GPIB Write and **GPIB Read**—found in **Functions>>Instrument I/O>>GPIB**. The
Help Windows for these VIs are shown here.

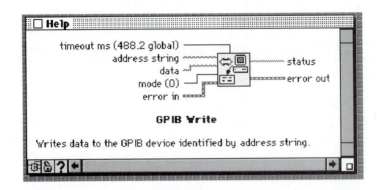

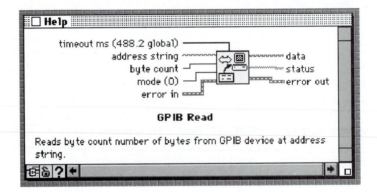

From these Help Windows, we find that to send a command to a particular instrument, one simply wires the correct ASCII character string to the **data** input of **GPIB
Write**. Upon execution, this icon will then use the GPIB to transmit this string to the
selected instrument. Data can also be retrieved from an instrument using **GPIB Read**,
which upon execution, presents the retrieved data at its **data** output. The Help
Windows also reveal that these two GPIB-related VIs have three configurational
inputs in common:

- **address string:** Because more than one instrument can be connected to the GPIB,
  the particular instrument of momentary interest is selected by wiring its identification number (in the form of a string) to this input. An instrument's identification number (*"GPIB address"*) can be any integer between 0 and 30 and is
  typically defined via a hardware DIP switch setting within the instrument or a
  sequence of button pressing and/or knob turning on its front panel. The instrument's user manual will describe the method for setting its address.

- **mode:** When a message is sent over the GPIB, some method for signaling its end-
  point is required. Several options for effecting this *termination* signal are avail-

able and your selection from among these various modes is indicated by an integer wired to the mode input. We will discuss these modes in detail below.

- **timeout ms:** If, for example, an instrument to which a string is to be written doesn't seem to be responsive (perhaps a nameless experimenter ineptly forgot to flip on the instrument's POWER switch—okay, I admit it, I've done this a couple of times), the VI will only attempt the requested GPIB operation for the number of milliseconds wired to this input. In such "communication breakdown" situations, the timeout feature prevents a data-taking program from being caught in an endless loop. If this input is left unwired, the icon defaults to a timeout period of 10,000 ms = 10 seconds.

A concrete example might sharpen your understanding of the GPIB scheme for computer-controlled experimentation as well as the utility of the two GPIB-related VIs. Consider a Tektronix TDS Digitizing Oscilloscope. According to its user's manual, if one writes the ASCII command string *CH1:VOLTS 1.00E + 0* to this instrument over the GPIB, it will respond by setting the y-axis scaling on its **Channel 1** trace to 1.0 Volt/Division. All the scope's other front-panel settings such as the x-axis scaling of Time/Division and the triggering options are GPIB-controllable using their own associated ASCII command strings, which are listed in the instrument's user manual. Additionally, the transfer of digitized waveform data from this instrument into a computer's memory can be accomplished by the following sequence of GPIB actions: First, perform a GPIB Write of the ASCII command *CURVE?* from the computer to the instrument. When the instrument receives this command, it loads the requested data into an output buffer that is part of its on-board GPIB interface circuitry. Once the data is loaded into the buffer, performing a GPIB Read will convey this data from the instrument to the computer's memory.

What hardware do you need to make possible this remote-controlled data taking in your own laboratory? First, you must have an appropriate National Instruments GPIB board plugged into an expansion slot of your computer. For Windows-based machines, this board might be the AT-GPIB or PCI-GPIB model, while Macintosh users might choose the NB-GPIB or PCI-GPIB board. Several stand-alone instruments, each equipped with GPIB circuitry, can then be connected to the plug-in board using special GPIB cables. When purchasing a stand-alone instrument, its GPIB capability is either included as a standard feature or as an (approximately $500) option. In the following exercises, I will assume that you have the required hardware system—computer with GPIB board, GPIB cable, and stand-alone instrument with GPIB interface. In addition, I will assume you have determined the instrument's GPIB address (in later versions of LabVIEW, the **Instrument Wizard** in the **Project** menu will assist you in this task).

In this chapter, you will learn how to write a LabVIEW VI that controls the GPIB-based communication between a stand-alone instrument and your computer. Because the GPIB communications scheme is so robust, you can actually write highly dependable data-taking programs with just the information already presented. However, with a bit more grounding in the GPIB system, you'll be able to create programs in which you can have near-total confidence. The following paragraphs will take you to the next level of sophistication in GPIB programming.

Features of the GPIB

When remote control of laboratory instruments first became possible, there was a chaotic period during which, more or less, each instrument manufacturer defined its own communications protocol through a unique blend of parallel and serial modes, positive and negative polarities, and assorted handshaking signals. In 1965, Hewlett Packard ended this cacophony by designing a universal instrument interface called the Hewlett-Packard Interface Bus (HP-IB) and offered it as the only option on all of its new computer-programmable instruments. Because of its high transfer rates, HP-IB quickly gained popularity and, in 1975, was accepted as an industry-wide standard known as IEEE-488. An improved version of this standard called IEEE-488.2 was adopted in 1987. The standard has expanded the application of HP-IB to instruments manufactured by companies other than Hewlett Packard, so everyone (except HP) now refers to IEEE-488 as the "General Purpose Interface Bus."

The GPIB system permits up to 15 instruments to be linked together via specialized 24-wire cables. Any connected device can be enabled as a *"talker"* (source of data) and any combination of remaining devices as *"listeners"* (recipients of data). A *"controller"* (usually the GPIB board plugged into your computer) dictates the role of each device. The data transfer occurs one byte at a time, coordinated by hardware handshaking, and can proceed at rates in excess of 1 Mbytes per second.

The 24-wire GPIB cable consists of 16 signal and 8 necessary grounding lines with the following organization:

- **Data Lines:** Eight of the signal lines, labeled DIO1-DIO8, are used for the parallel transfer of one-byte messages, whether data or commands, each bus cycle. All commands and most data use the 7-bit ASCII code set, in which case the DIO8 bit is unused or used for parity checking.

- **Handshake Lines:** The exchange of message bytes along the eight data lines is guaranteed to be without error by three "handshaking" lines labeled DAV (data valid), NRFD (not ready for data), and NDAC (not data accepted). The talker and listeners synchronize the transfer of a byte on the DIO lines by setting and clearing logic levels on the handshake lines. This process is exceedingly reliable, safely allowing an experimentalist to remain blissfully ignorant of its low-level details.

- **Interface Management Lines:** These five active-low logic level lines (that is, LOW equals TRUE and HIGH equals FALSE) manage the flow of information across the GPIB, indicating such things as the nature of a message byte being transferred (whether it is a command or datum), the completion of a message string, and the need for service from the controller. Here is a list of the functions of each of these management lines. Pay special attention to the operation of EOI and SRQ, as an understanding of these two lines will equip you to write properly functioning GPIB-based data-taking programs.

  - **ATN (attention):** This line is LOW when the controller is sending a command on the eight data lines and HIGH when the designated talker is sending data on the DIO lines.

  - **SRQ (service request):** Any device connected to the GPIB can "request service" from the controller by driving this line LOW. For example, an instrument can be pre-programmed to assert a service request once it has completed a data-gathering task, signaling the controller that it is ready to read out the measured values. After noting that this line is LOW, the controller identifies the particular instrument requesting service by performing a serial or parallel "polling" process. Both of these polling options will be described below.

- **EOI (end or identify):** This line can be used in two ways. First, a talker can drive this line LOW to indicate that the current byte on the DIO lines is the end of a message string. Second, in response to a service request, the controller can solicit a *parallel poll* by driving both the EOI and ATN line LOW simultaneously. Each instrument connected to the GPIB is associated with a particular DIO line and every device responds to the parallel poll by sending a one-bit status report (LOW if requesting service, HIGH otherwise) along its respective data line. By scrutinizing these status bits, the controller then determines which device is requesting service.
- **IFC (interface clear):** The controller drives this line LOW to initialize the bus into some known, predetermined state.
- **REN (remote enable):** When this line is driven LOW (or HIGH) by the controller, instruments on the bus can be controlled *remotely* using the GPIB (or *locally* using front-panel controls).

Common Commands

The IEEE 488.2 standard, which was adopted in 1987, enhanced and strengthened the GPIB by specifically defining methods for various interface tasks. One important such innovation was the introduction of a standardized set of *"common commands"* for the many generic operations that all GPIB instruments must perform. The mnemonics for these common commands begin with asterisks to delineate them from the other device-specific commands recognized by a particular instrument. All IEEE 488.2 compliant instruments, at the very least, are required to recognize the subset of 13 common commands given in the following table. Many of these commands are related to the reporting of events using two status registers called the SBR and SESR, which will be described in detail starting in the next paragraph.

Mandatory Command Commands	Function
*IDN?	Reports instrument identification string
*RST	Resets instrument to known state
*TST?	Performs self-test and reports results
*OPC	Sets operation complete (OPC) bit in SESR upon completion of command
*OPC?	Returns "1" to the output buffer upon completion of command
*WAI	Waits until all pending operations complete execution
*CLS	Clears status registers
*ESE	Enables event-recording bits in SESR
*ESE?	Reports enabled event-recording bits in SESR
*ESR?	Reports value of SESR
*SRE	Enables a SBR bit to assert the SRQ line
*SRE?	Reports SBR bits that are enabled to assert the SRQ line
*STB?	Reports the contents of the SBR

Status Reporting

Another IEEE 488.2 innovation is a standardized scheme for *status reporting*. This status report system is available to inform you of significant events that occur within each instrument connected to the GPIB. In this scheme, each instrument, as part of its GPIB interface, is equipped with two status registers called the *Standard Event Status Register (SESR)* and the *Status Byte Register (SBR)*. Each bit in these registers records a particular type of event that may occur while the instrument is in use such as an execution error or the completion of an operation. When the event of a given type occurs, the instrument sets the associated status register bit to a value of one, if that bit has previously been enabled (see following). Thus by reading the status registers, you can tell what type of events have transpired.

The Standard Event Status Register, which is schematically shown here, records eight types of events that can occur within a data-taking instrument.

Standard Event Status Register (SESR)

7	6	5	4	3	2	1	0
PON	URQ	CME	EXE	DDE	QYE	RQC	OPC

The eight events associated with the eight bits of the SESR are described in the following table. In our work, the OPC bit will be most useful.

Bit	Associated Events of SESR
7 (MSB)	**PON** (Power On): Instrument was powered off and on since the last time the event register was read or cleared.
6	**URQ** (User Request): Front-panel button was pressed.
5	**CME** (Command Error): Instrument received a command with improper syntax.
4	**EXE** (Execution Error): Error occurred while instrument was executing a command.
3	**DDE** (Device Error): Instrument is malfunctioning.
2	**QYE** (Query Error): Attempt was made to read the instrument's output buffer when no data was present, or a new command was received before previously requested data had been read from the output buffer.
1	**RQC** (Request Control): Instrument requests to be controller.
0 (LSB)	**OPC** (Operation Complete): All commands prior to and including an *OPC command have been executed.

The SESR exists as an event-signaling tool for you to use in your programs. However, this status register completely lacks initiative and will not perform any work unless you request it to do so. Thus, when initiating communications with an instrument, one of the messages that you may wish to send is an instruction that activates the subset of event-reporting SESR bits that are of interest to you. For instruments that conform to the IEEE 488.2 standard, this activation process is accomplished via the

ESE (Event Status Enable) command. For example, suppose you wish the QYE bit to be activated and thus record any execution errors in the SESR's bit 2. Since $00000100_2 = 4_{10}$, the QYE bit can be activated by performing a GPIB Write of the ASCII command *ESE 4* to the instrument. In our work to come, we will activate the OPC bit with the command *ESE 1*.

The Status Byte Register, which is schematically shown here, records whether data is available in the instrument's output buffer, whether the instrument requests service, and whether the SESR has recorded any events.

Status Byte Register (SBR)

7	6	5	4	3	2	1	0
—	RQS	ESB	MAV	—	—	—	—

The function of the eight bits of the SBR are described in the following table. The SBR bits are studious, performing their status reporting duties without need of a request from you.

Bit	Function of SBR Bit
7 (MSB)	May be defined for use by instrument manufacturer.
6	**RQS** (Request Service): The instrument has asserted the SRQ line because it requires service from the GPIB controller.
5	**ESB** (Event Status Bit): An event associated with an enabled SESR bit has occurred.
4	**MAV** (Message Available): Data is available in the instrument's output buffer.
3–0	May be defined for use by instrument manufacturer.

An instrument can be configured to assert the SRQ line in response to either of two events—an event detected by the Standard Event Status Register or the presence of previously requested data in the output buffer (that is, the assertion of the ESR or MAV bit, respectively). This configuration process is accomplished on IEEE 488.2 instruments by using the *SRE* (Service Request Enable) command. For example, if you wish an event detection by the SESR to trigger a request for service by the instrument, initialize the instrument by writing the ASCII command *SRE 32* to the instrument over the GPIB. Since $32_{10} = 00100000_2$, the setting of the SBR's fifth (ESB) bit will then be the criterion for the instrument asserting the SRQ line. If, instead, you wish the presence of data in the output buffer (signaled by the MAV bit HIGH) to trigger a SQR, then write *SRE 16*. Finally, *SRE 0* will disable the instrument's ability to assert the SRQ line.

The relationship between the Standard Event Status Register, the GPIB-related output buffer, and the Status Byte Register (along with the common commands that configure and query each) is illustrated in the following diagram.

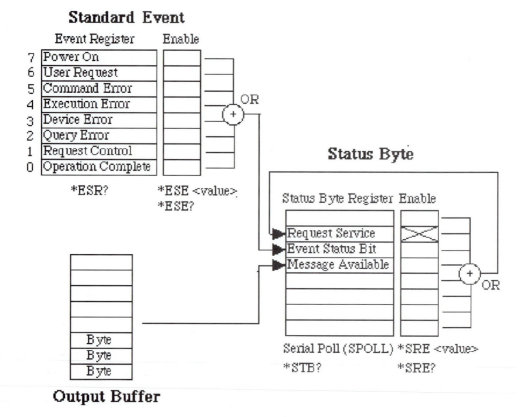

As an alternative to the use of the SRQ line, *serial polling* is a common method for determining the status of an instrument. In a serial poll process, the controller queries a device and the device responds by returning the value of the bits in its Status Byte Register. A serial poll is easily accomplished in LabVIEW using **GPIB Serial Poll**, found in **Functions>>Instrument I/O>>GPIB**. This icon's Help Window is shown here.

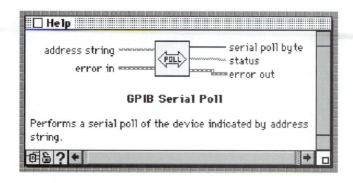

Device-Specific Commands

Finally, each stand-alone instrument is designed for a specialized purpose and has its own idiosyncratic methods for accomplishing its objectives. Thus, every programmable instrument comes with a set of "*device-specific*" commands that allow the user to control its functions remotely and to transfer the information it produces into a computer's memory. The array of device-specific commands for an instrument is listed in its user's manual. This set of commands is defined by the instrument's maker...and therein lies a problem. When surveying the user's manuals for programmable instruments of varying models and manufacturers, you will find a great diversity in the style of the various command sets. Some (especially those associated with older model instruments) are an alphabetized collection of cryptic one or two character strings (the designers' thinking was obviously "shorter commands yield quicker and, therefore, better computer-instrument communication"). At the other extreme are the user-friendly sets, with similar commands logically grouped, each represented by an easy-to-read-and-remember mnemonic.

As programmable instruments have come into wider use, it has become apparent that development costs and unscheduled delays can be diminished markedly by simplifying the instrument programmers' task whenever possible. Thus, user-friendly device-specific command sets are the rule, rather than the exception, for instruments being currently manufactured. Commonly, these command sets are organized in a hierarchical *tree structure*, similar to the file system used in computers. Each of an instrument's major functions, such as **TRIGger**, **MEASure** (alternately, **SENSe**), **CALCulate** or **DISPlay**, define a *root*, and all commands associated with that root form its *subsystem*. So, for example, to configure the Hewlett Packard 34401A Digital Multimeter to measure a DC voltage whose value is expected to fall within the range of ±10 Volts (an action within its **SENSe** subsystem), the appropriate command appears as follows:

SENSe:VOLTage:DC:RANGe 10

Here, *SENSe* is the root keyword and colons (:) represent the descent to the lower-level *VOLTage*, then *DC*, then lowest-level *RANGe* keywords. Finally, *10* is a parameter associated with *RANGe*. While the full command mnemonic given here can be sent to the instrument, it is only absolutely necessary to send the capitalized characters.

In 1990, a consortium of equipment manufacturers defined the *Standard Commands for Programmable Instruments (SCPI)* in an effort to standardize the device-specific command sets of computer-controlled instrumentation. While this standard has not been universally adopted, it is not uncommon to discover that your post-1990 instrument is SCPI-compliant. As a means of categorizing generally applicable command groups, the SCPI standard posits the following model for a generic programmable instrument. An instrument whose function is to perform measurements on an input signal is assumed to have the root functions shown in the next diagram. Here, for example, **SENSe** includes any action involved in the actual conversion of an incoming signal to internal data such as setting the range, resolution and integration time, while **INPut** consists of actions that condition the signal prior to its conversion such as filtering, biasing, and attenuation.

SIGNAL MEASUREMENT INSTRUMENT

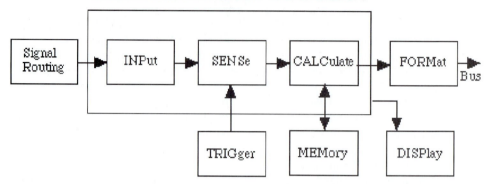

Alternately, an instrument that generates signals is modeled by the following diagram.

SIGNAL GENERATION INSTRUMENT

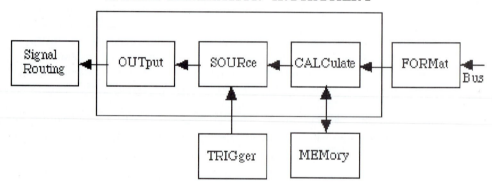

The SCPI command set is organized in a hierarchical tree structure using the syntax illustrated on the previous page by the HP34401A Multimeter command. You'll learn more about the SCPI command syntax as you work your way through this chapter. But maybe now is a good time to dive in and actually program an instrument.

SIMPLE GPIB WRITE/READ OPERATION

In designing this chapter, I faced the following problem. There are thousands of GPIB-interfaced stand-alone instruments available for purchase from the myriad of world-wide scientific instrument makers. I have a small subset of these instruments in my laboratory and I can use them there to practice the art of GPIB communication. You also, hopefully, have a small subset of such instruments available to practice with in your own laboratory. What's the problem? Well, because of the high cost and specialized nature of such equipment, the probability that my subset and your subset have some common instrument is most likely very small. The unfortunate thing about this situation is that each stand-alone instrument is designed to take specialized measurements and, for the most part, only understands its own unique set of GPIB commands (which are defined by its maker and are listed in its user's manual). Thus, before attempting to control a particular instrument using the GPIB, the programmer must have a detailed understanding of the measurement that that instrument is designed to take, the procedure that it implements in doing its work, and the command list that it recognizes. All of these considerations greatly constrain the writing of a set of generic laboratory exercises that everyone can perform.

That said, I still was faced with the fact that I had to choose a particular instrument to work with in this chapter's exercises. For the following reasons, I chose the Hewlett Packard 34401A Multimeter. First of all, the HP 34401A measures voltage, current, and resistance—vanilla-flavored quantities that require no specialized knowledge to understand (unlike, for instance, the control of grating angle and slit size in a spectrometer). Second, for such a high quality and useful GPIB-equipped instrument, its price tag of approximately $1000 makes it extremely affordable. Every lab should have one and many do! Third, this instrument is both IEEE 488.2 and SCPI compliant. Thus, the following exercises, rather than being narrowly tied to one specific device, can be much more universally applicable by demonstrating the generic features of these widely used standards such as command syntax and status reporting.

In the best circumstance, a HP34401A is already available for your use (or, with a modest investment, you can purchase this worthwhile instrument). Then, without need for modification, you can straightforwardly work your way through the given exercises to learn the basics of GPIB communication. If, instead, you have some other GPIB-equipped instrument available, try reading the following pages to understand the generic issues being investigated. Then, by consulting the user's manual, it may be fairly easy, for instance, to substitute an ASCII command string here and there, to adapt the exercises to your particular instrument. If neither of the above describes your situation, simply read through the following pages. I believe you will learn some valuable features of the GPIB that will serve you well in future work.

Using the GPIB Write and GPIB Read Icons

Let's begin by learning to implement the most fundamental of LabVIEW's GPIB-based VIs—**GPIB Write** and **GPIB Read**. These VIs are found in **Functions>>Instrument I/O>>GPIB**, the palette that contains the *"traditional GPIB"* functions. Traditional GPIB VIs are universally applicable, in that they are compatible with both (older) IEEE 488 and (newer) IEEE 4.88.2 compliant instruments and GPIB boards. The more specialized **Functions>>Instrument I/O>>GPIB 488.2** palette will be discussed in a few moments.

First, try writing the mandatory IEEE 488.2 common command *IDN?* to your instrument, the query that requests the instrument to return its identification string. Construct the following front panel with a single **String Indicator** labeled **Instrument Response**, which will display the identification string received from the instrument. You'll want to resize the indicator so that it can display a string much larger than its default size allows. Save this VI under the name **GPIB Simple Write/Read** in **YourName.llb**.

Switch to the block diagram and construct the following code within **frame 0** of a **Sequence Structure**. Wire a **String Constant** to the **data** input of **GPIB Write**, then enclose the ASCII command *IDN?* within it. When this command is sent to an IEEE 488.2 compliant instrument over the GPIB, the instrument responds by loading its identification string into the output buffer of its GPIB interface circuitry. Leave the **timeout** input unwired. The timeout period then defaults to 10 seconds.

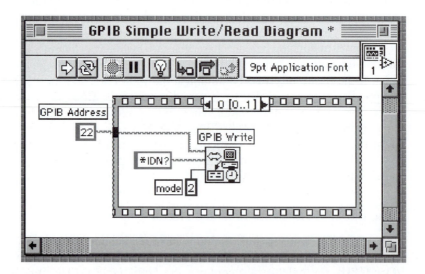

Remember that the integer labeled **GPIB Address** is enclosed within a **String Constant**, not a **Numeric Constant**. The string format is used for the **address string** input to accommodate device identification in complex situations (which we will not be investigating) where multiple controllers must be delineated and/or devices require primary as well as secondary names. Then plus signs (+) and colons (:) are used in the address name of a device, necessitating the string format. Code the correct address for your device on the diagram (in my case, it is *22*).

Termination of GPIB Write

At the conclusion of the process of writing a command to a device, some method must be used to signal that the complete message has been passed. The methods available for effecting this termination signal are listed in the following table, along with the integer code number that LabVIEW assigns to each. An instrument's user's manual will list the particular termination method that it recognizes in a GPIB Write process. The appropriate method can be selected by wiring its integer code number to the **mode** input of **GPIB Write**. In the table, LF denotes the ASCII linefeed character, also called newline (decimal 10, hexadecimal 0A). CR signifies the ASCII carriage return character (decimal 13, hexadecimal 0D).

Mode	Termination Method for GPIB Write
0	Send EOI with the last character of the string
1	Append CR to the string and send EOI with CR
2	Append LF to the string and send EOI with LF
3	Append CR LF to the string and send EOI with LF
4	Append CR to the string but do not send EOI
5	Append LF to the string but do not send EOI
6	Append CR LF to the string but do not send EOI
7	Do not send EOI

The IEEE 488.2 standard appoints LF as its special *end-of-string (EOS)* character. That is, when receiving a message string, the LF character is always interpreted by the receiver as the last byte of a message. Thus, appending LF to a command string (mode = 5) is one method of signaling message termination in IEEE 488.2 communication. Alternately, the IEEE 488.2 standard allows the assertion of the EOI line while the last character in the string is being passed (mode = 0) as another acceptable termination method. Finally, one can implement both of the above (mode = 2) to terminate a message. Since the HP34401A is IEEE 488.2 compliant, any of these possibilities will provide proper message termination. I chose the "overkill" method (mode = 2) in the previous block diagram.

After execution of **frame 0**, the instrument's identification string will be present in its GPIB-related output buffer. We now need to read the buffer's contents over the GPIB into the computer. Add **frame 1** to the **Sequence Structure** and wire the **GPIB Read** icon as shown next. When executed, this icon will read up to N bytes from the output buffer of the selected GPIB device, where N is equal to the integer wired to the **byte count** input. The HP34401A user's manual specifies that the instrument's identification string (including terminators) can be up to 35 characters long. Thus, wire the **byte count** input to an integer greater than or equal to 35.

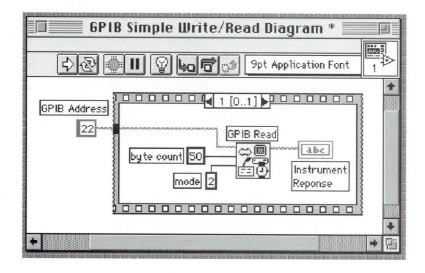

Termination of GPIB Read

When receiving the contents of the instrument's output buffer, your computer needs some method of discerning the completion of the message. The value of **byte count** offers one method of read termination. After the GPIB board within your computer reads **byte count** number of bytes, it stops the read process. Alternately, an instrument may designate an arbitrary character, typically LF or CR, as the EOS character. Then, whenever the GPIB board reads that special character, it knows that the current message is completely received. Or, finally, the instrument can assert the EOI line as a termination signal. LabVIEW assigns an integer code to each of these termination methods and, by wiring its code number to the mode input, you can select the particular method appropriate to your instrument. The following table lists these mode assignments.

Mode	Termination Method for GPIB Read
0	No EOS character. The EOS termination mode is disabled, so read terminated either by EOI or **byte count**.
1	EOS character is CR. Read terminated on either EOI, **byte count,** or CR (whichever comes first)
2	EOS character is LF. Read terminated on either EOI, **byte count,** or LF (whichever comes first).
x	The decimal number x defines the desired EOS character. Read terminated on either EOI, **byte count**, or x (whichever comes first).

As mentioned already, the IEEE 488.2 standard stipulates that the EOS character is LF, and that message transfers may be terminated with this EOS and/or by asserting the EOI with the message's last byte. This termination method corresponds to mode = 2 in the table. Since the HP 34401A is IEEE 488.2 compliant, the integer *2* is wired to the **mode** input of the **GPIB Read** icon in the previous block diagram.

Return to the front panel and run the VI. If all goes well, **Instrument Response** will display the instrument's identification string upon completion of the VI execution, as shown.

According to the HP34401A user's manual, this identification string identifies the instrument's manufacturer and model number followed by some integers that denote the version numbers of installed firmware that controls three internal microprocessors.

GPIB Simple Write/Read will leave the multimeter in *remote mode* with its triggering circuitry "idled." You can return to *local mode*, which continuously "triggers" measurements, by pressing the instrument's front-panel SHIFT/LOCAL key.

Performing a Measurement over the GPIB

Now that we have a template for the GPIB write-then-read process, let's try controlling a real measurement. Hook up some known DC voltage difference, say 5 or 6 Volts, between the HI and LO Voltage Inputs of the HP34401A Multimeter. The HP34401A user's manual instructs us that delivering the following sequence of ASCII commands to the instrument over the GPIB will result in one DC voltage measurement being taken, then loaded into the device's output buffer:

CONF:VOLT:DC<Space>10,0.00001

INIT

FETC?

Here is the meaning of this secret code. First, the HP34401A can be programmed to perform 11 different types of measurements including DC voltage, AC voltage, DC current, AC current, resistance and frequency. Given these options, the first command instructs the instrument that we desire to take a DC voltage measurement. The full command is *CONFigure:VOLTage:DC <Space><Range>,<Resolution>* (this command actually executes a collection of commands drawn from the HP34401A's **INPut**, **SENSe**, **TRIGger**, and **CALCulate** root subsystems). The command mnemonic *CONFigure:VOLTage:DC* is constructed in the hierarchical path structure, typical of SCPI-compliant instruments. *CONFigure* is the root-level keyword and colons (:) represent the descent to the lower-level *VOLTage*, then lowest-level *DC* keywords. While the full command mnemonic can be sent to the instrument, it is only absolutely necessary to send the capitalized characters. Separated from the command mnemonic *CONF: VOLT:DC* by a *<Space>*, the numerical values for two measurement parameters— *<Range>* and *<Resolution>*—are specified. *<Range>* selects among the instrument's five available voltage measurement scales. Each scale offers a different sensitivity with *<Range>* giving the maximum measurable value on a particular scale. The five available ranges are 100 mV, 1 V, 10 V, 100 V, and 1000 V. In our situation of measuring a signal of approximately 5 Volts, the 10 V scale is appropriate. *<Resolution>* specifies the precision of the measurement, with the options of three levels of accuracy—$4\frac{1}{2}$, $5\frac{1}{2}$ and $6\frac{1}{2}$ digits (the $\frac{1}{2}$ digit means that the most significant decimal place can only take on a value of "1" or "0"). Thus on the 10 V scale, voltages can either be resolved at the level of 0.001, 0.0001, or 0.00001 Volts. The trade-off in requesting higher accuracy is that the measurement takes a longer time. In the command sequence above, the highest resolution of $6\frac{1}{2}$ digits is selected by setting *<Resolution>* equal to *0.00001* when *<Range>* equals *10*. Note the syntax of the *CONF* command, which obeys the conventions of the SCPI language: a *comma* (,) separates the parameters from each other and a *<Space>* separates the mnemonic from the parameters.

Once the multimeter has been configured for the desired measurement as described in the previous paragraph, the data-taking process is begun by sending the *INITiate* command (from the **TRIGger** root subsystem). Upon receipt of *INIT*, the multimeter will acquire the requested voltage reading, then store this value in its internal memory. Finally, the *FETCh?* command (from the **MEMory** root subsystem) instructs the instrument to transfer the reading in its internal memory to its GPIB-related output buffer.

We'd like now to place this command sequence into **GPIB Simple Write/Read**. Since there are three commands to be sent, it appears that we must modify the VI to include a sequence of three successive implementations of **GPIB Write**. While you are free to do so, a much easier solution is available. The SCPI language allows the programmer to concatenate several commands together into one long multi-command string that can be sent in a single **GPIB Write** statement. The syntax for this concatenation process is as follows:

- Use a semicolon (;) to separate two commands within the string.
- Begin a command with a colon (:) if it has a different root-level than the command preceding it. The first command in the concatenated string and IEEE 488.2 common commands (which begin with an asterisk) do not require a leading colon.

Since each of our three commands has a different root-level, applying these rules results in the following concatenated string:

CONF:VOLT:DC<Space>10,0.00001;:INIT;:FETC?

Modify **frame 0** in the block diagram of **GPIB Simple Write/Read** as shown next.

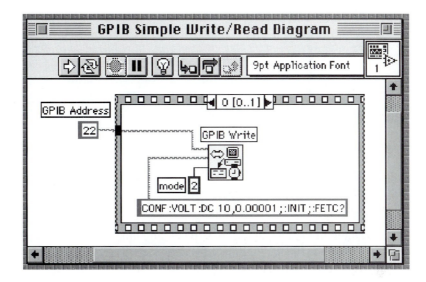

Run the VI. Your computer will instruct the multimeter to acquire a $6\frac{1}{2}$ digit voltage reading, retrieve this value, then display it on the front panel. Cool, eh?

Note that, while extra digits are displayed, the value within **Instrument Response** is only accurate to the fifth decimal place.

As shown, the HP34401A reports its data measurements in the form of an ASCII character string using the exponential format SD.DDDDDDDDESDD, where S = a positive or negative sign, D = a numeric digit, and E = exponent. For future reference, note that the string that represents a data value is 15 bytes long. If you wanted to use this reading as input to a mathematical calculation (a common situation), you would need to convert the string representation into a numerical format. Such conversion operations can be easily accomplished in LabVIEW using the array of conversion VIs found in **Functions>>String**. In the present case, use **From Exponential/Fract/Eng** in **Functions>>String>>Additional String To Number Functions**. The Help Window for this VI is given next.

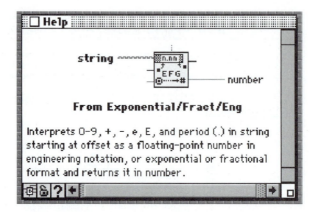

Place a **Digital Indicator** on the front panel and label it **Numeric Voltage**. Then modify **frame 1** on the block diagram as follows.

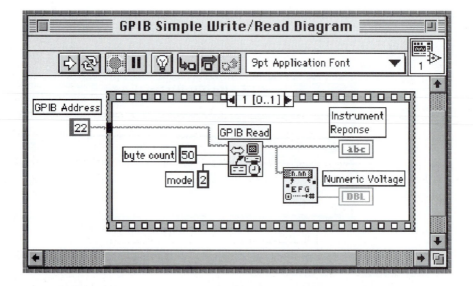

Run the VI to verify that the conversion VI performs as expected.

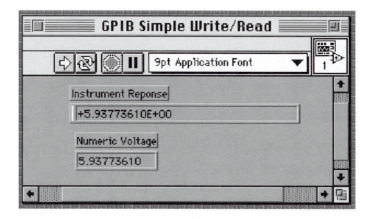

Using the IEEE-488.2 Send and Receive Icons

Finally, with the intent of writing programs that can be successfully reproduced by the maximum number of readers, the traditional GPIB VIs have been used in the previous block diagrams. However, if both the National Instruments GPIB board within your computer as well as the stand-alone instrument with which you are communicating are IEEE 488.2 compliant, then you might consider using the VIs found in **Functions>> Instrument I/O>>GPIB 488.2**. Take a quick glance at this palette and you will discover that it offers a wide assortment of VIs which, when compared with the traditional palette, enhance your control over the GPIB. In the present case of controlling simple write and read operations, there is little to be gained in using the 488.2 palette. For practice, though, you might be interested in rewriting **GPIB Simple Write/Read** using **Send** and **Receive** from **Functions>>Instrument I/O>>GPIB 488.2**. The Help Window for **Send** is shown below.

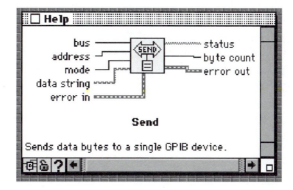

Here, the **bus** input requests the GPIB board's identification number. Unless you have more than one such board plugged into your computer, you may leave **bus** unwired as the board's ID number is the default value of *0*. The GPIB address of the stand-alone instrument to which you wish to send **data string** should be wired to the integer (**I16**) **address** input. **mode** is an integer (**U8**) code that selects the desired termination method from the IEEE 488.2 approved possibilities as given in the following table.

Mode	Temination Method for Send
0	Do nothing to mark the end of the transfer
1	Append LF to the string and send EOI with LF
2	Send EOI with the last character of the string

The Help Window for **Receive** appears as follows.

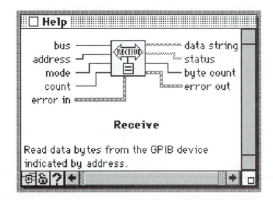

The **bus** input may again be left unwired in most cases, while **data string** is the ASCII message read over the GPIB from the stand-alone instrument identified by the integer (**I16**) **address**. The integer (**I32**) **count** defines the maximum number of ASCII characters expected for the **data string** message. If you wish to terminate the read operation using an end-of-string character, use the integer (**U16**) **mode** input to define the desired EOS character. For the common EOS choice of LF (whose decimal equivalent is 10), wire a **Numeric Constant** enclosing the value of *10* to **mode**. The read operation then terminates when either **count** bytes are read, EOI is asserted, or the EOS character is detected, whichever comes first.

Use **Save As**… to make a copy of **GPIB Simple Write/Read** with the new name **GPIB Simple Write/Read 488.2**. Then code the following block diagram.

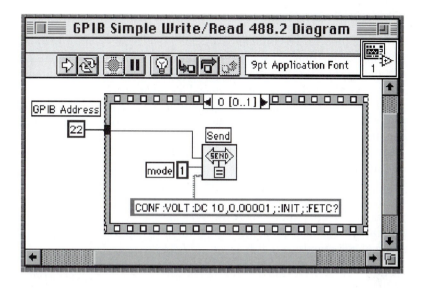

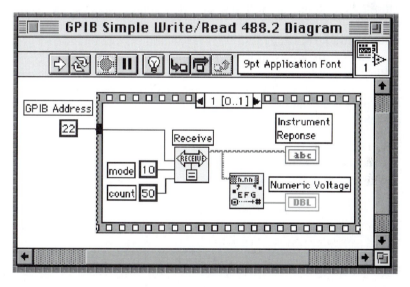

Run **GPIB Simple Write/Read 488.2**. If it doesn't succeed in obtaining a voltage reading from the HP34401A, possibly your GPIB board is not IEEE 488.2 compliant. Once you're finished, save your work in **YourName.llb**.

SYNCHRONIZATION METHODS

Although most GPIB commands are completed quickly after being received by a programmable instrument, some commands start a process that requires a significant amount of time (such as acquiring a large amount of data or moving an object from Point A to Point B). The time required for such processes must be taken into account when writing a data acquisition program, else, upon execution, the program may request data before it is available, induce undesirable motion, or cause some other chaotic outcome.

As an example, in its default configuration, the HP34401A multimeter makes one measurement after receipt of the *INITiate* command, then stores the measured value in its internal memory. However, through use of the *SAMPle:COUNt* *<Space><Value>* command, the multimeter can be instructed to take and store multiple measurements upon receiving *INITiate*. The HP34401A is configured to acquire 100 DC voltage measurements with $6\frac{1}{2}$-digit resolution via the following concatenated string of commands:

CONF:VOLT:DC<Space>10,0.00001;:SAMP:COUN<Space>100;:INIT;:FETC?

The *FETCh?* command will load the 100 acquired values from the multimeter's internal memory (which, by the way, can hold up to a maximum number of 512 measured values) into the instrument's GPIB-related output buffer.

Let's write a VI that uses the given command string to gather a sequence of 100 voltage measurements. Open **GPIB Simple Write/Read**, then use **Save As**... to create a new VI called **GPIB Write/Read (Long Delay).** Delete **Numeric Voltage** from the front panel and enlarge **Instrument Response** so that it can display a very long string (which is the concatenation of 100 voltage values).

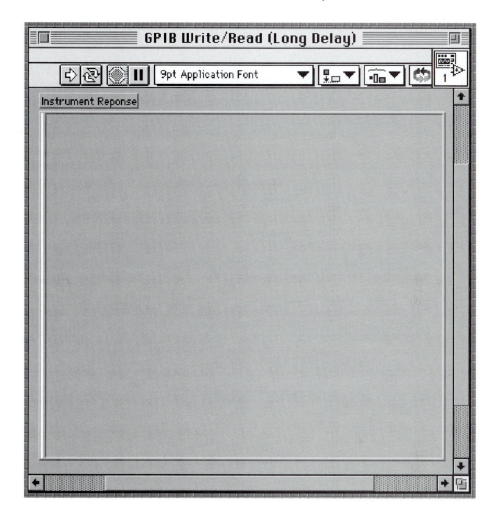

Modify the block diagram as shown in the following illustration so that the given command string is written to the multimeter in **frame 0**, then a string consisting of the 100 voltage measurements is read into the computer and displayed on the front panel in **frame 1**. Since each reading is 15 bytes long and a delimiting character will be needed to separate each reading, this string is expected to be about $(100 \times 15) + 100 = 1600$ bytes long. Input an integer larger than 1600 to **byte count**.

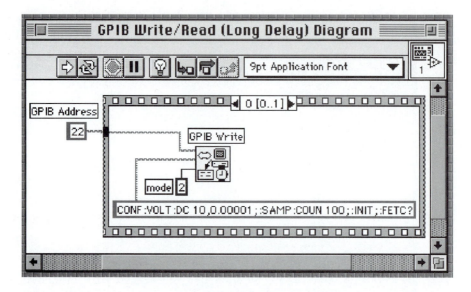

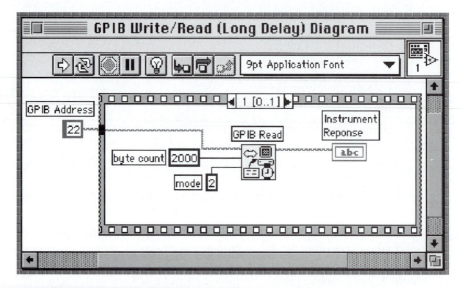

Run **GPIB Write/Read (Long Delay)**. Count down the seconds 10...9...8... 7...6...5...4...3...2...1... Disappointed? Upon completion, you will find that your VI will not display even one voltage measurement (let alone the expected 100 values). After some head scratching and checking of the HP34401A user's manual, the following explanation emerges for the failure. Simply stated, voltage measurements take time. In its default configuration, it takes the multimeter 10 power-line cycles (PLC) for each voltage measurement. Additionally, the HP34401A has an autozero feature, which is enabled by default. This feature operates as follows: After each measurement, the multimeter internally disconnects the input signal and takes a zero reading. The instrument then subtracts the zero reading from the preceding measured value to prevent offset voltages in the multimeter's internal circuitry from affecting measurement accuracy. Since the zero reading also takes 10 PLC, each complete voltage measurement by the HP34401A takes 20 PLC. Assuming the multimeter is plugged into a 60 Hz power source (that is, 60 PLC per second), 100 voltage measurements will take about

$$100\left(\frac{20\ \text{PLC}}{60\ \text{PLC/sec}}\right) = 33.3\ \text{seconds}$$

There's the problem! Remember **GPIB Read** "times out" after 10 seconds, but the measurement we have initiated takes over 30 seconds. Thus long before the requested data is available, **GPIB Read** terminates the execution of your VI.

There are a couple of crude solutions to our dilemma. First, you can insert an intermediate frame in the block diagram's **Sequence Structure** that simply contains a **Wait (ms)** icon, wired to produce a delay of about 33 seconds between the issuance of the data-taking command and the order to read the gathered data values. Or, for a slightly more elegant fix, you can change the **timeout** value for **GPIB Read** from its default value of 10000 ms to something larger than 33 seconds as shown here. Make this modification in your block diagram.

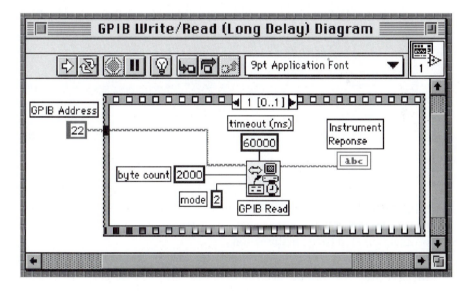

Now run **GPIB Write/Read (Long Delay)**. About 33 seconds later you should see something like the following front panel. Note that the delimiter used by the HP34401A to separate neighboring data values is a comma.

In the preceding example, we found that with a detailed knowledge of the measurement process being implemented, it is possible to troubleshoot a malfunctioning GPIB-based VI. Please note that lack of communication (in particular, the GPIB board not correctly knowing when the instrument's data will be available) is the root problem that led to the malfunction.

Fortunately, powerful tools exist that allow one to monitor the status of tasks being performed by a programmable instrument. For IEEE 488.2-compliant instruments, these tools are the Standard Event Status Register (SESR) and Status Byte Register (SBR) that were discussed at the beginning of this chapter. With proper use of the SESR and SBR, many potential data-taking glitches, such as the one just experienced, can be avoided.

The status reporting capabilities of the SESR and SBR can be employed in several ways. We will explore two commonly used techniques—the Serial Poll and Service Request Methods. The core operation for both of these methods is the same—the completion of an assigned task triggers the Operation Complete (OPC) bit in the Standard

Event Status Register to be set, which in turn sets the Event Status Bit (ESB) of the Status Byte Register.

In the Serial Poll Method, the setting of ESB is detected by directly checking the Status Byte Register, whose state is obtained by serial polling the instrument. The complete step-by-step process of this method is shown in the following diagram.

Serial Poll Method

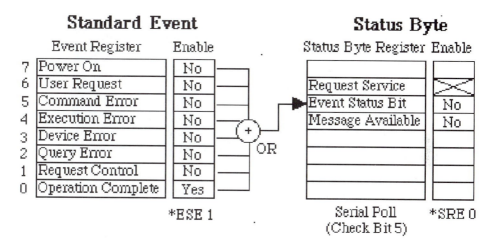

In the Service Request Method, the Status Byte Register is configured such that, when its ESB is set, the Request Service bit is induced to be set also. This action then causes the instrument to assert the SRQ line of the GPIB, which alerts the controller that the assigned operation is complete. This method is pictured here.

Service Request Method

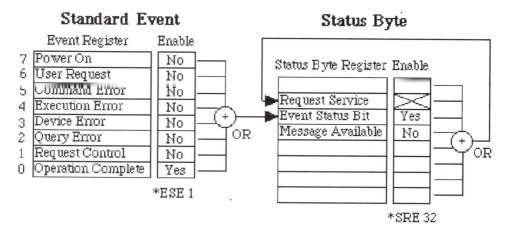

We'll write VIs that implement both of these approaches to status reporting.

Measurement VI Based on the Serial Poll Method

Let's try the Serial Poll Method first. To configure the HP34401A for status reporting using the Serial Poll Method, write the following VI called **Status Config (Serial Poll)** and save it in **YourName.llb**. A suggested front panel and terminal assignments are shown.

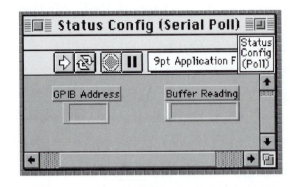

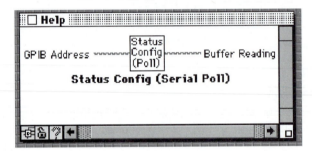

The block diagram consists of the following three-frame **Sequence Structure**.

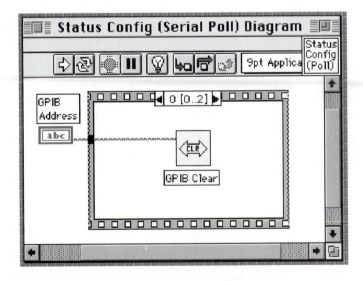

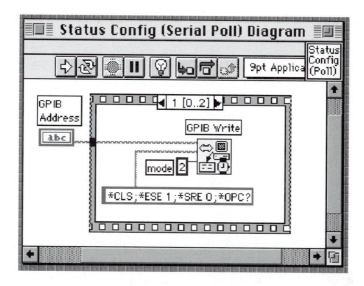

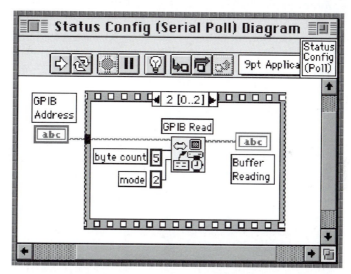

Here's how the VI works. In **frame 0**, we execute **GPIB Clear** (found in **Functions>>Instrument I/O>>GPIB**), whose Help Window is shown below.

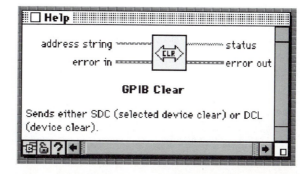

Although not an absolute necessity for inclusion in **Status Config (Serial Poll)**, this VI performs the precautionary action of "clearing" the HP34401A. **GPIB Clear** instructs the multimeter to abort all measurements in progress, disable its triggering circuitry, clear its GPIB-related output buffer and prepare to accept a new command string.

Next, **frame 1** sends the concatenated command string *CLS;*ESE 1;*SRE 0; *OPC? to configure the HP34401A for status reporting using the Serial Poll Method. Note that since the component strings are all IEEE 488.2 common commands, leading colons are not required in the concatenation. In this sequence of commons, *CLS clears the contents of the SESR and SBR. As described in the beginning of this chapter, *ESE 1 enables the SESR's OPC bit to set the ESB in the Status Byte Register and *SRE 0 disables the instrument from asserting the SRQ line. Then, *OPC? requests the instrument to return a "1" to the GPIB output buffer after this command is completed. This last command is included simply as a method of checking that the entire sequence of commands has been executed.

Finally, **frame 2** reads the contents of the GPIB output buffer. If all goes well, there should be a single ASCII character 1 read into the computer.

Run **Status Config (Serial Poll)**. Upon completion, does the string indicator **Buffer Reading** display an ASCII character 1?

Next, construct a VI called **Serial Poll** that continuously reads the Status Byte Register of an instrument until a given bit is set. A suggested **Serial Poll** is shown here.

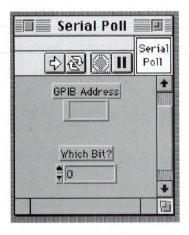
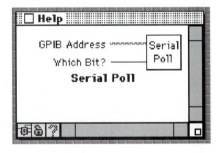

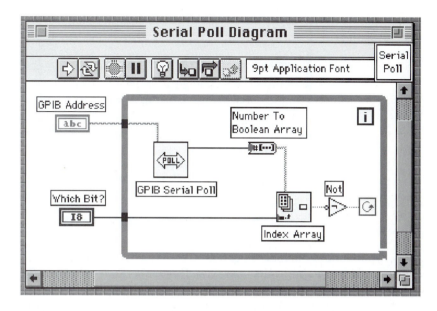

GPIB Serial Poll, found in **Functions>>Instrument I/O>>GPIB**, is the workhorse of this VI. With each iteration of the While Loop, its **serial poll byte** output returns the current values of the SBR's eight bits in the form of an integer. For example, if the SBR's fifth bit (ESB) is set, then **serial poll byte** outputs the integer *32*, since $00100000_2 = 32_{10}$. The Help window for **GPIB Serial Poll** is shown here.

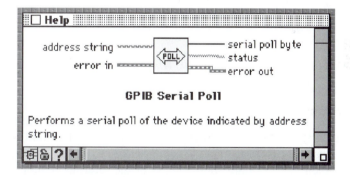

The individual bits of **serial poll byte** can be checked through the use of **Number To Boolean Array** (found in **Functions>>Boolean** with its Help Window given next). This VI creates an array of TRUE and FALSE values that mirror the sequence of zeros and ones (starting from the least-significant bit) in the binary representation of the integer input **number**. For example, if **number** equals the decimal integer *48*, then the **Boolean array** output will be [F, F, F, F, T, T, F, F], since $48_{10} = 00110000_2$. **Index Array** can then be used to ascertain the value of a particular element in this array. With the use of **Not** (from **Functions>>Boolean**), **Serial Poll**'s While Loop will continue to iterate until the **Which Bit?** bit becomes TRUE.

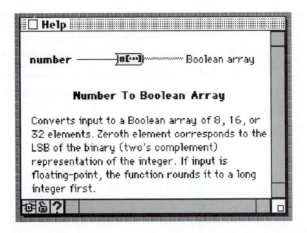

Run **Serial Poll** under **Highlight Execution** and, through your observations, gain a better understanding of its operation. Remember to input values for **GPIB Address** and **Which Bit?** on the front panel. When run in this isolated manner, the VI will most likely never be able to exit the While Loop, so you'll have to stop it using the **Abort Execution** button in the toolbar.

We're finally ready to write **GPIB Write/Read (Serial Poll)**. This master program implements serial polling to synchronize the GPIB activities necessary in acquiring 100 voltage measurements using a HP34401A Multimeter.

Open **GPIB Write/Read (Long Delay)**, then use **Save As**... to create **GPIB Write/Read (Serial Poll)**. The front panel can remain unchanged. Switch to the block diagram and modify it as shown below, with **Status Config (Serial Poll)** and **Serial Poll** used as subVIs in **frame 0** and **frame 2**, respectively. The concatenated command string

*CONF:VOLT:DC<Space>10,0.00001;:SAMP:COUN<Space>100;:INIT;*OPC;:FETC?*

sent to the instrument in **frame 1** works as follows: After configuring the multimeter for the desired sequence of DC voltage measurements, the acquisition process is begun by the *INIT* command. The secession of 100 values is determined and temporarily stored in the HP34401A's internal memory. After the hundredth value is acquired, **OPC* instructs the instrument to set its SESR's OPC bit (which, in turn, sets the SBR's ESB), then *FETC?* loads the contents of the internal memory into the GPIB output buffer. In **frame 2**, **Serial Poll** detects the setting of ESB, which then allows the instrument's GPIB output buffer to be read by **GPIB Read** in **frame 3**. One might be tempted to write this command with **OPC* after *FETC?*, rather than sandwiched between *INIT* and *FETC?*, as above. It is best, however, to avoid sending **OPC* after a query (a query is a command like *FETC?* that ends in a question mark) as such commands cause a message to be loaded into an instrument's GPIB output buffer. If the message exceeds the finite size of the output buffer, as happens in our present situation, the query must be immediately followed by **GPIB Read** as the program executes in order to read the long message string over the bus successfully.

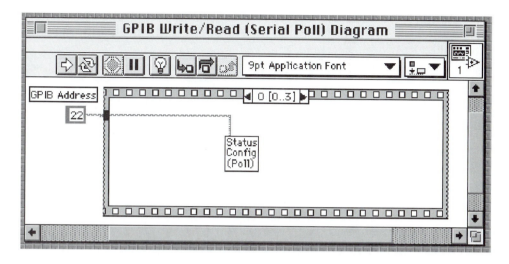

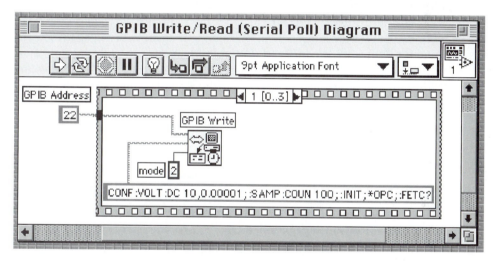

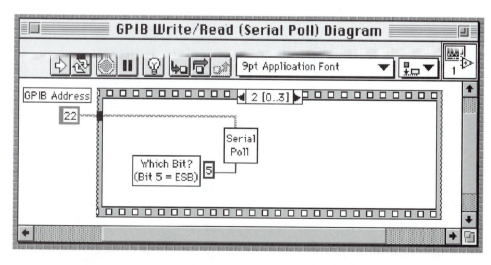

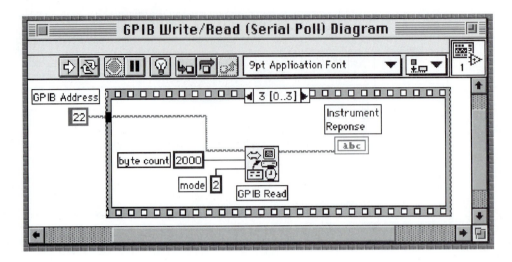

Run **GPIB Write/Read (Serial Poll)**. Does the VI obtain the requested 100 DC voltage measurements successfully? If so, try running it again with **Highlight Execution** activated for both **GPIB Write/Read (Serial Poll)** and its subVI **Serial Poll**. This exercise will illustrate the weakness of the Serial Poll Method, namely, the large volume of bus traffic required by this technique. During the 30-odd seconds while the 100 data values are being gathered, the instrument is polled countless times by the controller so its status can be continuously monitored. While effective, the Serial Poll Method is rather inefficient because of its excessive use of the bus and processor time.

Measurement VI Based on the Service Request Method

The Service Request Method provides status reporting with a minimum of GPIB activity. To configure the HP34401A for status reporting using the Service Request Method, open **Status Config (Serial Poll)**, then create **Status Config (SRQ)** using **Save As…** and save it in **YourName.llb**. A suggested front panel and terminal assignments are shown here.

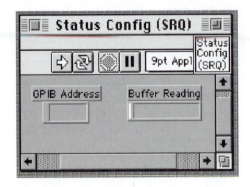

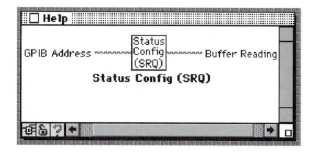

Only one modification of the block diagram is needed. As shown next, by changing *SRE 0* to *SRE 32* in the command string sent to the instrument, the HP34401A will assert the SRQ line when the SBR's fifth (ESB) bit is set. The already present *ESE 1* command configures the instrument to set the ESB in response to the SESR's OPC (Operation Complete) bit being set.

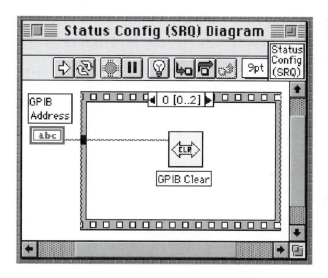

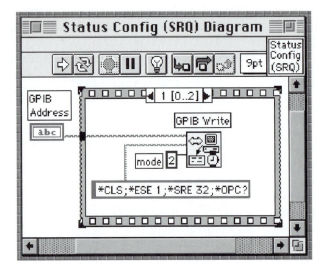

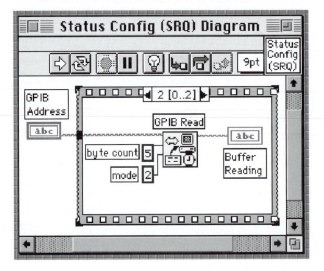

Save your work.

Open **GPIB Write/Read (Serial Poll)**, then use **Save As**... to create **GPIB Write/Read (SRQ)**. The front panel is fine as is. Switch to the block diagram and modify it as shown next.

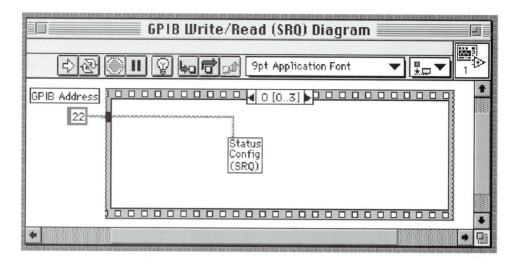

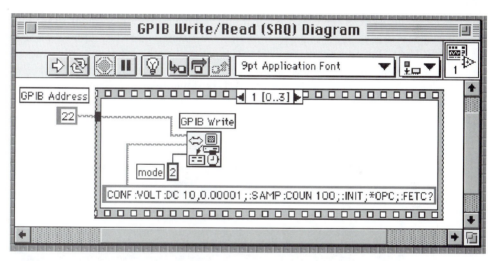

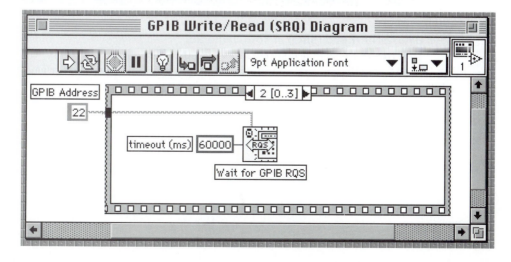

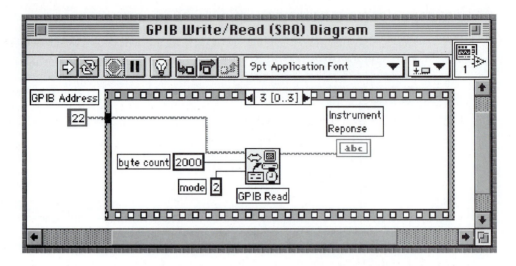

The VI **Wait for GPIB RQS**, found in **Functions>>Instrument I/O>>GPIB** (Help Window shown below), sits idly until the instrument denoted by **address string** asserts the SRQ line. However, there is a limit to the patience of this VI. It will only wait up to a total time of **timeout ms**, with a default value of 10,000 milliseconds = 10 seconds. Because our measurement requires over 33 seconds, a constant larger than 33,000 must be wired to the **timeout ms** input of **Wait for GPIB RQS**, as shown in the block diagram above.

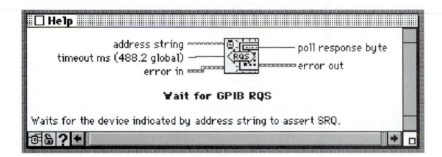

Save **GPIB Write/Read (SRQ)** in **YourName.llb.**

Run the VI and see if it successfully acquires the requested 100 DC voltage values. Do you understand the operation of this program and how the Service Request Method manages to work with a minimum of GPIB activity?

CREATING AN INSTRUMENT DRIVER

An instrument driver is a collection of modular software routines that perform the operations required in the computer control of a particular programmable instrument. These operations include configuring, triggering, status checking, sending data to, and receiving data from the instrument. **Status Config (Serial Poll)** and **Status Config (SRQ)** are examples of configuration VIs that would be useful to include as part of the HP34401A software driver. You will now write another configuration VI, this time, one that prepares the multimeter for taking a desired measurement.

The HP34401A is capable of implementing 11 types of measurements: DC and AC voltage, DC voltage ratio (ratio of voltage at two different inputs), DC and AC current, 2- and 4-wire resistance (2-wire is the "normal" method for measuring resistance; the more involved 4-wire technique is necessary only when measuring very small resistance samples), frequency and period of an AC signal, continuity, and diode check. To gain experience with some of the LabVIEW tools available for developing instrument drivers, let's write a VI called **Measurement Config** that offers the choice of configuring the HP34401A for either a DC voltage, AC voltage, or 2-wire resistance measurement. You, of course, can be more ambitious and write your VI to control up to all 11 possible measurement types.

Referring to the HP34401A user's manual, one finds that **Measurement Config** should be programmed with the following list of three possible commands. The user then selects the appropriate command from this menu to configure the instrument for the desired measurement type.

CONFigure:VOLTage:DC <Space> <Range>, <Resolution>

CONFigure:VOLTage:AC <Space> <Range>, <Resolution>

CONFigure:RESistance <Space> <Range>, <Resolution>

Also, the manual instructs that the possible values of <Range> for both the DC and AC voltage measurements are 0.1, 1, 10, 100, and 1000 Volts. For the resistance measurement, the allowed <Range> values are 100, 1k, 10k, 100k, 1M, 10M, and 100M ohms. In all cases, the measurement precision may be $4\frac{1}{2}$, $5\frac{1}{2}$, or $6\frac{1}{2}$ digits, which corresponds to <Resolution> being $10^{-4}, 10^{-5}$, or 10^{-6} times the <Range> value, respectively.

Create a VI named **Measurement Config** and save it in **YourName.llb**. The **Function** and **Resolution Enumerated Type** controls have three entries each— *DC Voltage, AC Voltage, Resistance* and $4\frac{1}{2}$ *Digits*, $5\frac{1}{2}$ *Digits*, $6\frac{1}{2}$ *Digits*, respectively. **Range (DBL)** and **Sample Count (I16)** are **Digital Controls**, while **Autozero** is a **Labelled Square Button** and **GPIB Address** is a **String Control**.

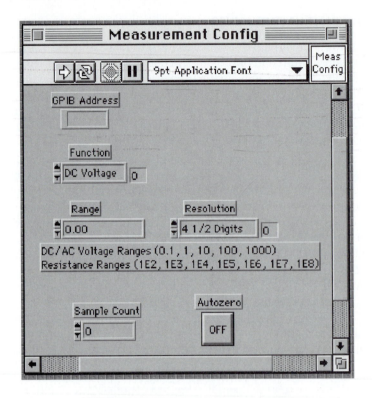

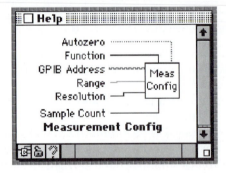

Switch to the block diagram and code it as shown in the followng illustration.

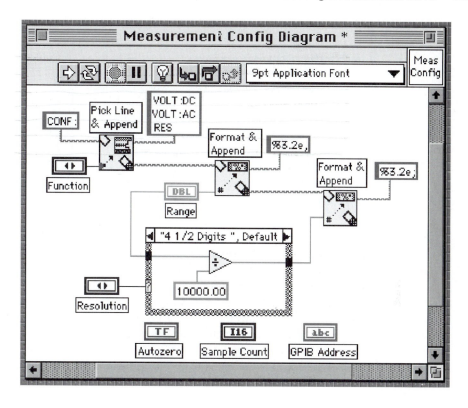

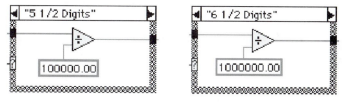

This diagram constructs the desired command string in a three-step process. First, all three possible commands begin with the keyword *CONF:*, so this sequence of ASCII characters is wired to the **string** input of **Pick Line & Append** (found in **Functions>>String** with Help Window shown below). The value of the **line index** input (given by the front-panel **Function Enumerated Type** control) then selects which of the three possible lines programmed into the **multiline string** is to be appended to *CONF:*. Note, Windows users will create the three lines in **multi-line string** by the following sequence of keystrokes: *VOLT:DC<Space> <Enter>VOLT:AC<Space> <Enter> RES<Space>*. Macintosh users, the sequence is the same except use *<Return>* instead of *<Enter>*. Be sure to include the *<Space>* character at the end of each command string.

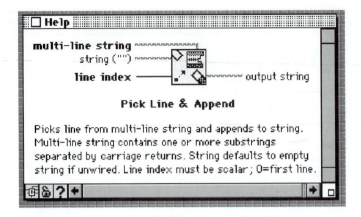

Format & Append, from **Functions>>String>>Additional String to Number Functions** (Help Window shown next), then is used to attach two more string fragments, each with imbedded ASCII-coded numbers, which program the *<Range>* and *<Resolution>* settings of the multimeter. This icon takes the value at the **number** input, converts it to an ASCII representation, which is included with a defined format as part of the character sequence given at the **format string** input. **format string** is appended to **string** and presented at **output string**. A percent symbol (%) indicates the location that the ASCII representation of **number** is to be inserted in **format string**, and it is followed by the formatting information (see page 123). In the previous diagram, the scientific notation format *3.2e* is used for both *<Range>* and *<Resolution>* parameters. Note a comma (,) and semicolon (;) follow *<Range>* and *<Resolution>*, respectively.

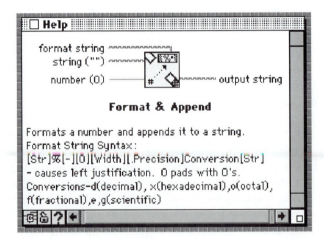

Finally, include code to program the desired number of measurements to be taken following an *INITiate* command and to control the multimeter's autozeroing feature. The *%5d* format in the *SAMPle:COUNt* command specifies a five-place decimal integer. The format string entry for this command should be *:SAMP:COUN<Space>%5d*.

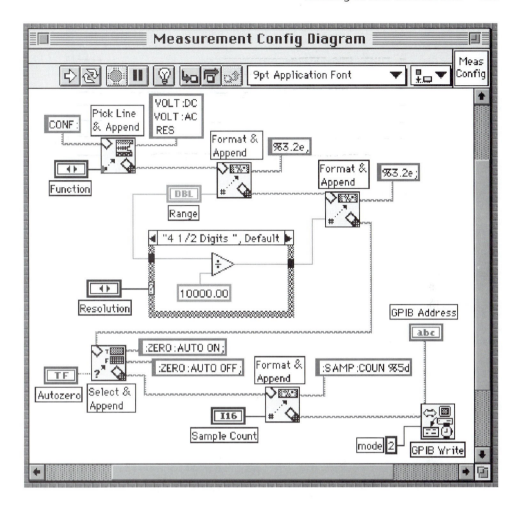

Autozero can be either turned on or off with the following commands:

ZERO:AUTO<*space*>ON
ZERO:AUTO<*space*>OFF

Select & Append, found in **Functions>>String** with the following Help Window, provides an easy way to choose which of these two choices is concatenated to our command string. Remember to include the leading colon and final semicolon in the **false string** and **true string** entries to assure proper command concatenation.

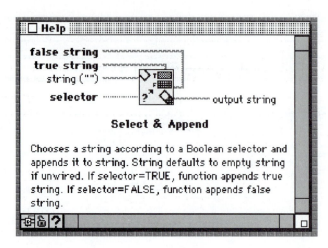

To guarantee that the instrument fully processes the sent command string before exiting this VI, conclude the string with *OPC?*. Then, in a succeeding **Sequence Structure** frame, use **GPIB Read** to obtain the value *"1"* from the GPIB output buffer, signaling the command's completion. The required additions to the block diagram are shown next.

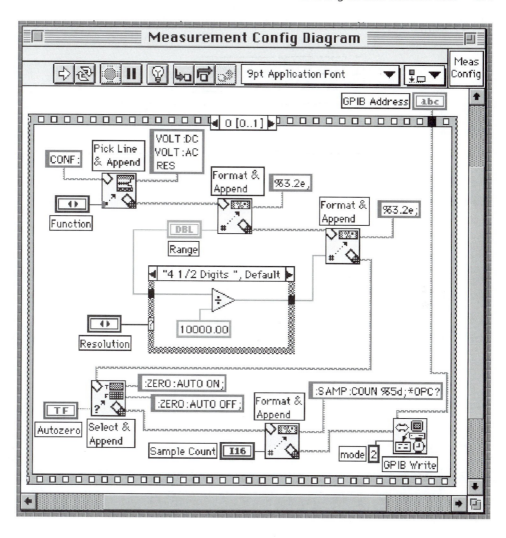

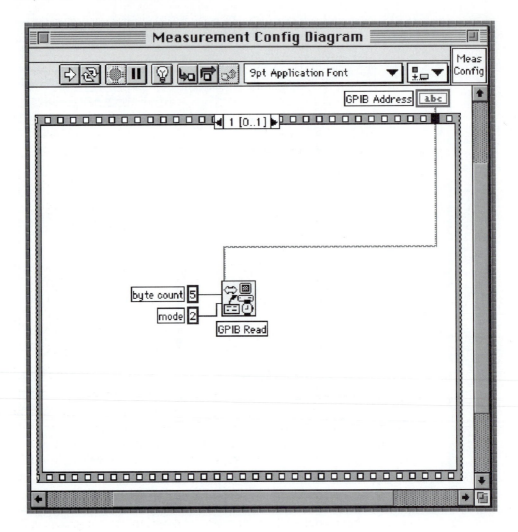

Try running **Measurement Config** to see if it properly configures the multimeter. If the instrument beeps, indicating an error in the sent command, place a Probe on the string being passed to the input of **GPIB Write** and check that the concatenated command has a form such as the following:

CONF:VOLT:DC<Space>1.00E+1,1.00E−5;:ZERO:AUTO<Space>ON;:SAMP: COUN<Space>5;*OPC?

Make sure all of the colons, semicolons, and spaces are included. This command string results from the following front-panel settings.

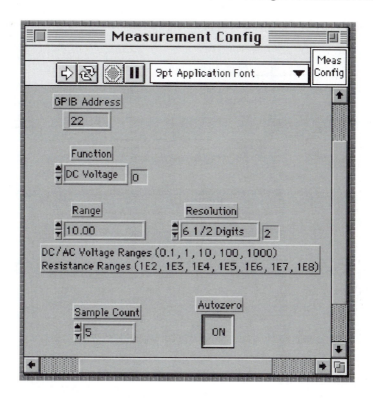

Measurement Config will leave the multimeter in *remote mode*. You can switch to *local mode* by pressing the instrument's front-panel SHIFT/LOCAL key. The HP34401A can then be triggered (equivalent to sending the INIT command over the GPIB) with the SINGLE/TRIG button. A star (*) annunciator will blink on the instrument's front-panel display during each measurement. Does this annunciator blink **Sample Count** times after the SINGLE/TRIG button is depressed?

Save your work on **Measurement Config**.

Write another modular VI for your HP324401A instrument driver called **Take Data** as shown below, and save it in **YourName.llb**. The leading *CLS* command assures that all bits in the SESR and SBR register are set to zero, prior to each data-taking process.

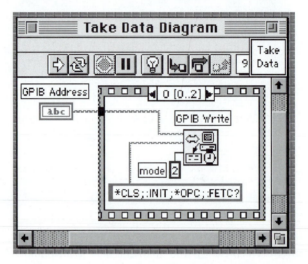

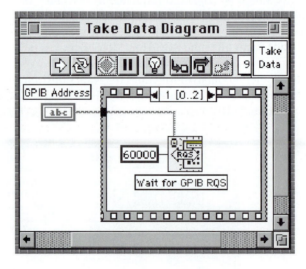

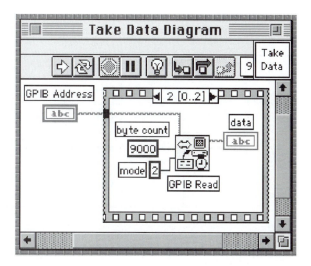

Open **GPIB Write/Read (SRQ)** and create **GPIB Write/Read (SRQ) 2** using **Save As**... Rewrite the block diagram using your modular driver software.

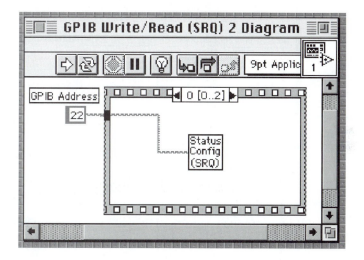

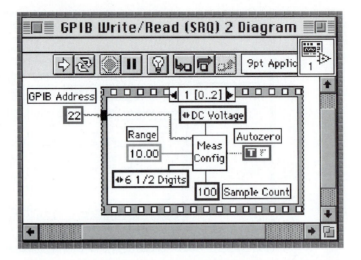

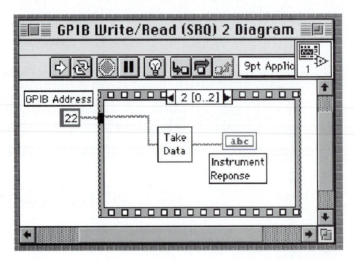

A useful utility to include in the HP34401A instrument driver performs the following task: Take the instrument's data string (data values delimited by commas and terminated by a LF character) and convert it to a numeric array and/or a spreadsheet format. One manifestation of that utility called **Re-Format Data String** is shown next.

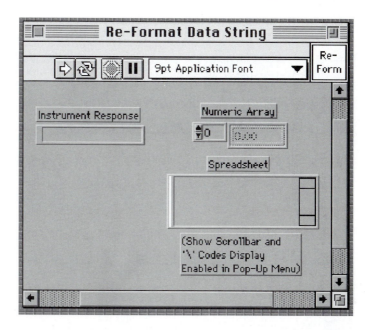

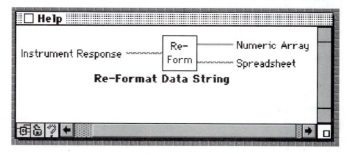

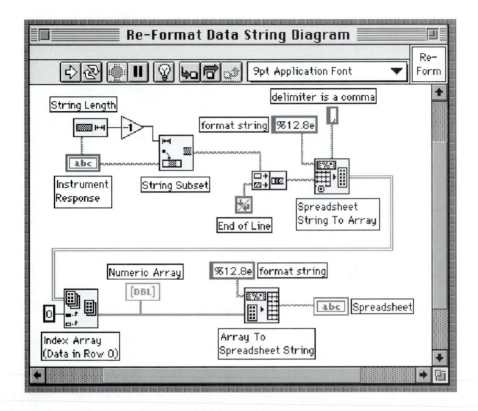

If interested, try writing **Re-Format Data String.** It implements many of the string icons found in the **Functions>>String** palette. Do you understand how it works? Once written, include this program as a subVI in **GPIB Write/Read (SRQ) 2** and watch it perform its magic.

Now that you know some of what goes into writing an instrument driver, here's some very good news. In many cases, the LabVIEW instrument driver you will need for a particular instrument in your laboratory has already been written and is available for your use free of charge. National Instruments provides an extensive library of instrument drivers as part of the Full Development System LabVIEW software package. These drivers can also be downloaded free of charge from *http://www.ni.com/idnet/*. You will find that most of these drivers are written using VISA (Virtual Interface Software Architecture) icons, which are generic instrument control VIs designed to communicate over three types of interfaces—GPIB, serial, and VXI. With the understanding that the VISA palette (found in **Functions>>Instrument I/O>>VISA**) has analogs to each icon in the GPIB palette (for example, **VISA Read** corresponds to **GPIB Read**), you should be able to understand the operation of these instrument drivers after your work in this chapter. In fact, LabVIEW comes equipped with the VISA driver for the HP34401A multimeter in **Functions>>Instrument Drivers**. Take a look at some of these subVIs and see if you can decipher them.

Using the Instrument Driver to Write an Application Program

Ultimately, the merit of an instrument driver is measured by the ease with which you can use it to write an *application program* to fulfill some specialized need in your scientific work. For sake of argument, let's say that you had a widget in your laboratory that was providing you with some interesting information about X, where X might be the temperature of an object or the intensity of a light source or the amount of time left before lunch. Additionally, say, the widget provides this information about X in the form of a "voltage code," that is, it produces an output voltage V that is some known function of X. Then, if you needed to monitor X (via measurement of V) as a function of time, you could do so using a HP34401A Multimeter and an appropriately written application VI. Such an application program, which implements the instrument driver subVIs you have developed, is given here. This VI, called **Time-Evolution of X**, provides real-time graphing of the *Voltage versus Time* data as it is being acquired and also provides the option of storing the data in a spreadsheet file. A **Waveform Graph**, as opposed to a **Waveform Chart**, is used for plotting so that the *time*-axis can be properly calibrated. Try building this program and running it to observe a time-dependent voltage input to the multimeter.

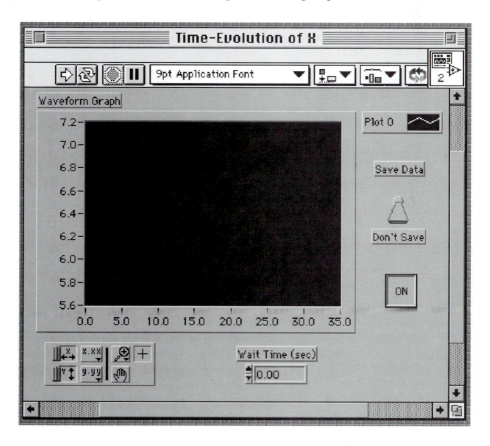

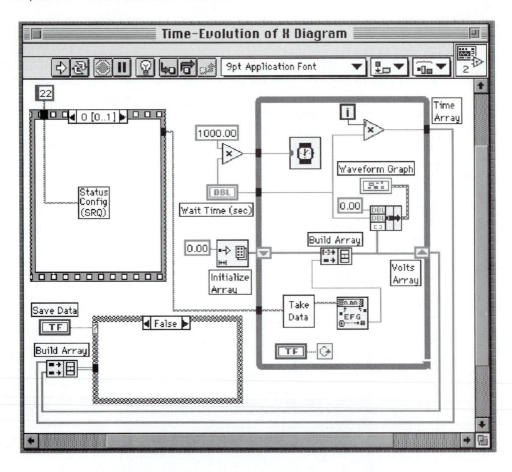

The bottom-left **Build Array** icon creates a 2D array from two 1D arrays at its inputs. If you get a bad wire when connecting to it, here's a hint for obtaining good wiring: One of the two inputs to **Build Array** isn't a 1D array because of a problem back at the While Loop.

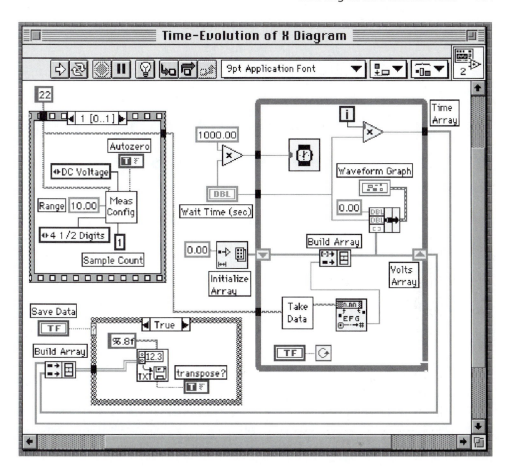

Construction of Temperature Control System

To perform the exercises in Chapter 11, you will need access to an apparatus that controls the temperature of a small object through the use of a thermoelectric (TE) device. There are, of course, a multitude of ways to construct the required gadget. In this appendix, I offer a design (total cost approximately $50) that has worked successfully for me.

First, as the object whose temperature is to be controlled, I use a rectangular aluminum block of dimensions $2" \times 1.5" \times 5/16"(1" = 2.54$ cm). Hereafter, I'll simply call this object "the block." So that accurate temperature measurements can be taken on the block, a small hole of diameter 3/16" is drilled into one of its sides to a depth of approximately 1/2". A thermistor (preferably coated with thermal grease, see following) can then be inserted into this hole and held securely by the addition of a 1/4" long #6-32 nylon set screw. For the thermistor, I use an inexpensive 10 kΩ model from Radio Shack (Part Number 271-110, $2) with soldered-on 12" lead wires for easy connection to a constant-current circuit. Finally, two #8-32 clearance holes are drilled (using #19 drill bit) into the block as shown in the diagram. The 1.450" spacing is chosen so that a 30-cm wide TE module will fit between these holes.

SIDE VIEW

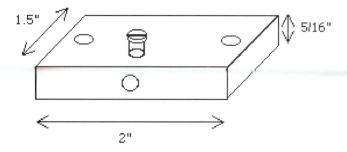

1.5"   5/16"

2"

TOP VIEW

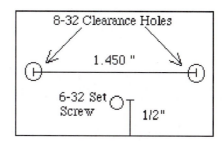

The required heating or cooling of the block is facilitated by placing it in contact with one side of a thermoelectric module. A TE module is a compact solid-state device that, via the Peltier Effect, acts as a heat pump. When current is caused to flow through the TE device in one direction, heat will be absorbed from the contacting block, cooling it, as shown in the following illustration.

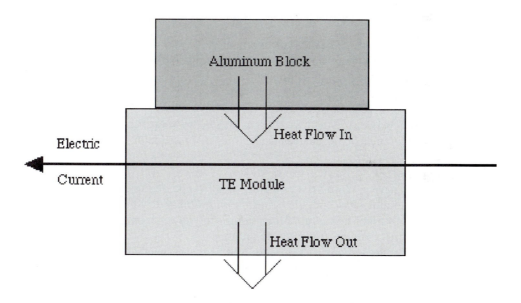

When current is made to flow through the TE module in the opposite direction, heat is pumped into the block, heating it.

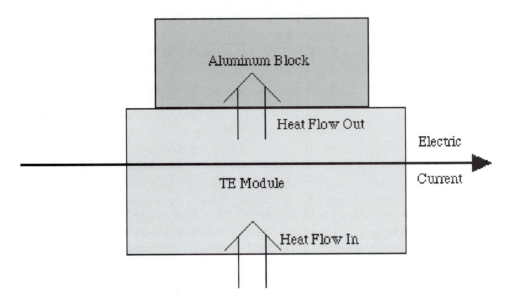

Note that, because the TE device is a heat pump, a not-pictured heat reservoir must be in contact with the bottom side of the TE module in both of the previous pictures. With insignificant change to its temperature, this heat reservoir accepts (provides) the heat pumped through the TE module when the block is being cooled (heated).

Melcor (1040 Spruce Street, Trenton, NJ 08648 USA, Telephone: 609-393-4178, FAX: 609-393-9461, World Wide Web–http://www.melcor.com) manufactures over 200 standard models of TE devices. The choice of a particular TE model is dictated by the heat-pumping capacity necessary and electric power supply limitations in your application. For our project, I use a Melcor Model CP 1.0-127-05L (approximately $15 each) TE device. Within the relatively small range of current flow between ±2 Amps, this TE module has enough heat-pumping ability to change the temperature of a 2" × 1.5" × 5/16" aluminum block significantly away from room temperature (including cooling it to below 0°C and icing it up!).

How can we construct the required heat reservoir? Remember that a heat reservoir is simply an object that can accept or produce an amount of heat (up to a maximum value determined by your particular application) without significantly changing its temperature. A big slab of aluminum (for example, 5" × 3" × 3") possesses a large "thermal mass" and thus can perform as a fairly decent heat reservoir. By tapping two #8-32 screw holes 1.450" apart into such a slab, a TE module can be sandwiched between the block and the slab. Then using two 3/4" long #8-32 (insulating) nylon screws, you can secure the three items together and, in the process, obtain good thermal contact between contacting surfaces. To guarantee efficient heat flow, coat the TE device's heat-pumping surfaces with thermal grease (available, for example, from Melcor, Jameco), before sandwiching the three items. This method for realizing a heat reservoir is inexpensive, but imperfect. If the TE module is run over moderately long time periods, the temperature of the aluminum slab will begin to significantly change from room temperature due to its rather poor efficiency at transferring heat with surrounding room.

A much more stable heat reservoir can be constructed using a finned aluminum heat sink. Thermalloy (2021 W. Valley View Lane, Dallas, TX 75234 USA, Telephone: 214-243-4321, FAX: 214-241-4656, World Wide Web–http://www.thermalloy.com) offers a line of products called *extrusions*, which are plate-like pieces of aluminum with many seamlessly attached fins. Due to the large surface area of its fins, an extrusion will very efficiently exchange heat with its surrounding room air, imbuing it with the desired properties of a heat reservoir. I have had good success using 6" lengths of Thermalloy Model 14468 Extrusion ($60 per 3 foot) as the heat reservoir in the temperature control system of Chapter 11. Tapping two #8-32 screw threads 1.450" apart into the plate-like surface of the extrusion, the TE module with the block atop can be mounted, then secured with two #8-32 (insulating) nylon screws. Again, use of thermal grease will enhance the conductance of heat between the block and heat reservoir. A final improvement results by mounting the extrusion on top of a small fan (for example, 120 VAC 4" Cooling Fan, Radio Shack Model 273-242, $17), which forces air flow through the extrusion's array of fins. Legs for the fan and small angle brackets to connect it securely to the extrusion can be fashioned from 1/16" aluminum strips. Attach these strips to the fan using the fan's mounting holes. A photograph of the fan with its homemade legs and brackets is shown here.

Photos (from two different angles) of the entirely assembled temperature control system follow. A small aluminum bar attached to the plate region of the extrusion acts as a strain-relief for the TE module wires.

The bi-directional current driver circuit for the TE module is shown next. For simplicity, the TIP transistors are represented in this diagram as simple bipolar transistors, but are, in actual fact, Darlington transistors. In regard to external connections, a Darlington transistor behaves exactly the same as to a single bipolar transistor with a very large current gain β (hence, the acceptability of the simplified representation). The Darlington's large β (on the order of 1000) results from its two-transistor internal construction, in which the collector current of one transistor provides the base current to the other. In its ON state, the Darlington's base-emitter voltage is "two diode-drops" ($\approx$1.2V). Because of the coupling between its two internal transistors, the Darlington is especially susceptible to thermal instabilities, necessitating proper heatsinking of these components for reliable service.

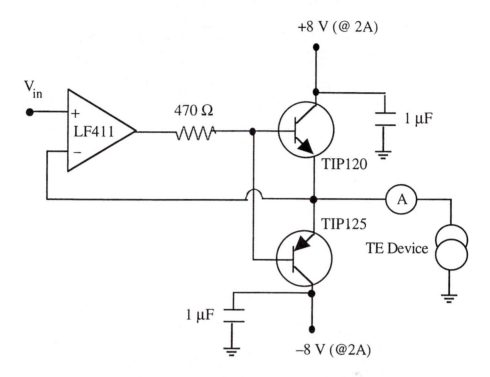

The $\pm$8 Volts power sources attached to the TIP transistor's collectors must be capable of providing up to about 2 Amps. The Laboratory DC Power Supplies sold by several electronic equipment manufacturers (for example, Tektronix PS280 or BK Precision 1760), which contain two identical DC supplies, are an ideal instrument for providing both the negative and positive voltage levels. These Laboratory DC Supplies typically have built-in ammeters which, in this circuit, can be used to monitor the current flow through the TE device. The 1 μF capacitors in the diagram remove any high-frequency noise present in the power supply voltage levels.

When this circuit is in operation, the TIP power transistors generate a lot of heat. These hot transistors present the following problems: (1) if you touch them, they will hurt your fingers (an avoidable danger, once you're aware of it), (2) if the circuit is set up on a solderless breadboard, they will melt the breadboard's plastic (this is the voice of experience talking) and (3) if the transistors are allowed to get too hot, they will become thermally unstable and possibly destroy themselves (again, the voice of experience). Proper heatsinking of the transistors will solve these problems. Because this process is somewhat involved, I provide students with a widget that contains the two transistors mounted on a heat sink. The noise-suppressing 1 μF capacitors are included as well. A top and bottom view of the widget (total cost approximately $30) is shown in the following photographs.

The two TIP transistors have TO-220 packaging. However, I mount them on a TO-3 Heat Sink (Jameco Part Number 16512, $4) from Jameco Electronics (1355 Shoreway Road, Belmont, CA 94002 USA, Telephone (800) 831-4242, FAX: 800-237-6948, World Wide Web–http://www.jameco.com). The pattern of mounting holes (appropriate for a single TO-3 packaged power transistor) on this finned heat sink allows the two TIP transistors to be affixed side by side as well as provides convenient openings to pass wiring through. A TO-220 Heat Sink Mounting Kit (Jameco Part Number 34121, $4 per five kits) must be used in mounting each TIP transistor to insulate its collector from the heat sink. The heat sink is then attached to a homemade chassis, fashioned from

1/16" aluminum sheet metal. Additionally, to facilitate necessary electrical connections, banana jacks and binding posts are mounted on the chassis.

The widget is then hardwired according to the following diagram. All of the parts in this diagram may be purchased from Newark Electronics (call 800-463-9275 or see http://www.newark.com to locate your local sales office) or from most any other electronics parts distributors.

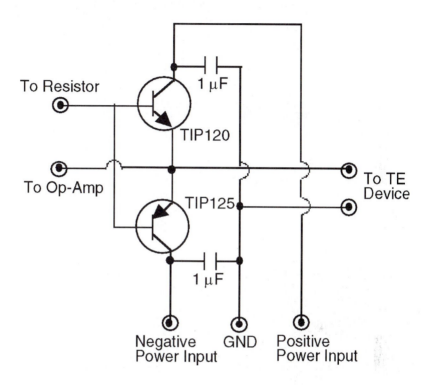

INDEX